The Real Inside... to Military Basic Training

A Recruit's Guide of Important Secrets and Hints to Successfully Complete Boot Camp

Revised Edition

Written by a recruit for a recruit. This is the most valuable guide you can ow... g.

The Real Insider's Guide to Military Basic Training: A Recruit's Guide of Important Secrets and Hints to Successfully Complete Boot Camp

Universal Publishers/uPUBLISH.com
USA • 2003

ISBN: 1-58112-597-6

www.uPUBLISH.com/books/thompson.htm

Important Notice: Some ideas expressed in this book are secrets drill sergeants do not want you to know. Several people in the military will not want you reading this guide. In fact, there are dishonorable individuals that have tried to criticize this publication by making false statements about it—especially on the internet. Do not believe the critics or others with hidden motives. Gather as much information as you can about the military, trust your instincts and make your own judgments.

For additional copies of this book, contact the publisher at:
www.upublish.com or www.upublish.com/books/thompson.htm

Acknowledgments

To my best friend and wife,
Lynne.

To Brooke, my daughter.

To all people in the military
protecting our freedom.

Thanks!

Table of Contents

NOTES

This page is intentionally left blank so the reader can make personal notes:

Introduction

Setting the Tone and Special Notes

Do you think the military is a good career choice for you? Do you have the physical and mental fortitude to complete basic training? To be sure, the military is an outstanding challenge and occupational opportunity. However, while most people are successful in the military, it is *not* for everyone. Assuming you are seriously considering enlisting in the armed forces, the fact that you are reading this book is an excellent first step. The more knowledge you have regarding military life, the more informed you are when making the crucial decision to enlist. Knowledge is power, so I implore you to use as many resources as you can before signing your enlistment contract.

Do not rely solely on this book to base your decision. My primary advice to people entertaining the idea of joining the military is to utilize prior service personnel, career counselors, and yes, even recruiters. Unfortunately, there are not many good books available. (I suggest referencing the Appendix of this book for recommendations). Spend the money and time researching the armed services to discover if it is a "good fit" with your personality and interests. I hope that this guide will help the reader understand his/her interests and make the best decision possible.

A good author always knows his audience and writes for a specific type of reader. Therefore, it is important that interested consumers understand the target market for this book. The information contained within the following pages is best suited for young people (17-30 years old) enlisting in the military—primarily in the Marines and Army, but Navy and Airforce personnel can find the information very useful. Most new enlistees are fresh out of high school; thus this guide is tailored accordingly. Additionally, my focus is on those readers enlisting at the lower pay grades of E-1 to E-4 (Private to Corporal, or Specialist) and below. Although *any* person entering the military will derive some benefit from this guide, officer candidates and college graduates should definitely consult additional resources.

The style of this book is purposefully written in a straightforward manner with just a hint of humor on a few pages. The guide is designed to be brief and easy to read. I am not trying to win a Pulitzer Prize; rather I am sincerely trying to pass on advice which will benefit new enlistees—advice I wish I had when experiencing basic training. As a recruit, time is not on your side and you will appreciate the brevity and simplicity of this handbook. However, be cautioned that some concepts and suggestions are repeated. Rest assured that anything which seems redundant is my way of emphasizing extremely important ideas. The reader will note that the summary at the end of some chapters efficiently underscores and recounts the primary points of that chapter. If you only read the chapter summaries every night before shipping out, you will have secured many suggestions without re-reading the entire guide.

Another point which some people might find unusual is that I mention things which appear to be common sense. Nevertheless, I know from experience that it is not wise to assume all people are aware of what constitutes common sense. Military recruits are a highly diverse set of people with equally diverse educational and intellectual abilities. What may seem evident to some readers is not so apparent to others. For example, some ill-informed recruits believe basic training is like a long camping trip and they have plenty of free time! Many recruits erroneously think everything they receive during training is free, but in fact they are required to pay for their haircuts (a shaved head), toiletries, running shoes, and other items.

If the reader has enlisted already, time is of the essence. Usually, new enlistees will have at least few weeks before shipping out. Start memorizing the parts of this book which are highlighted in the summary sections, start exercising seriously and continue to gather all the information you can about basic training. Underline key parts of this guide and write important notes within its intentionally wide margins. Remember, it is always easier to learn and memorize vital information in a non-stressful environment than it is during the chaotic conditions of training. The stress-induced environment of basic training wreaks havoc with one's ability to learn even simple tasks.

A personal note to the reader: When I entered the military a few years ago, there were no books specifically written about basic training. I relied on information given to me by a friend, my brother Sean (who was prior service) and an Army recruiter with vested interests. Due to the amazing lack of information about the military, I made it a point to take notes during my basic training. My intention to help other recruits by writing a guide actually gave me the drive to make it through the hardest months of my life. Currently, there are still only a few unbiased sources of information expressly regarding basic training. The books which are on the market appear to be good solid books. I will not criticize other authors' work if their purpose to help others is sincere.

However, my guide is a little different from the others because it is framed mostly from the standpoint of a new recruit, not an officer, drill sergeant, or recruiter. This is my second improved edition. In keeping with my mantra, "Written by a recruit for a recruit," I have conducted the most recent research on the latest in military training. Therefore, the reader benefits from having information given by recent military graduates and from an author who wrote about basic training as it was actually unfolding. I know what you will go through during basic because I not only experienced it as a recruit, I documented it when it was still fresh.

As a counselor and a school psychologist, my life is dedicated to the service of helping people realize their potential. To this end, I not only offer this book, but also give an e-mail address so others may correspond with me. Although I make no promises to answer all e-mails, I will honestly attempt to respond to inquiries.

As a precaution against liability, I also have to note that any comment I make is not given as professional advice, but only as my opinion. The readers must assume full responsibility for their actions. Use common sense, do not break any rules, and base your decisions on multiple sources.

If you have questions and want to hear my opinion, please e-mail me at my publisher's webpage www.upublish.com/books/thompson/htm. Again, I may not be able to answer due to the volume of inquiries, but I will try. Good luck and remember that you have already made a very smart move by investing in information.

To Navy, Marine and Air Force Recruits: As noted, Army personnel will derive the most benefit from this book. However, a good portion of the information is general enough to be helpful to any military recruit. For example, the physical conditioning regimen advice, tips regarding avoiding trouble, and understanding the general structure of military training are obviously useful to most enlistees. The aforementioned readers are best served if they read with an eye for broad information.

Generally speaking, the Marines will have to do more in terms of physical conditioning and combat arms training. The Navy and Air Force will focus more on service specific tasks and military discipline codes, and less on physical training than the Army, or Marines. However, the common factor within all military branches is that Initial Entry Training (IET, or basic training) is primarily focused on *indoctrinating* recruits into a radically different lifestyle. Therefore, recruits in other branches of the military might not acquire the specific information their Army counterparts do, but they should obtain a good general feel of what to expect during training.

Another point concerns the variations in training cycles. *Every* basic training class is slightly different and you will have a different experience than other recruits. I distinctly remember meeting another platoon of trainees midway through my training cycle. This other group told our platoon about their experiences on the far side of the base. In short, it appeared our neighboring platoon had a somewhat easier and partially different basic training experience. Drill instructors and company commanders have some latitude in how they train their recruits. Training tasks are dependent on the weather, availability of resources, and other variables. It seems that the Department of Defense delegates the general outline for all training cycles, but gives each base freedom on how to meet these broad requirements. Major training requirements for each branch of the military are outlined in the Initial Entry Training (IET) book given to each recruit.

Another factor which makes basic training experiences different is the personality traits and attitudes of individual recruits. Each recruit perceives and experiences his/her environment differently from each other. Some people have a very high tolerance to stress, while others become stressed out quite easily. Additionally, some people learn and comprehend military tasks quicker than others. Not everyone is the gung-ho, high- speed Sergeant Rock type. Individual differences certainly impact how basic training is viewed by various recruits and also accounts for the diversity in training cycles. As elementary as this sounds, recruits with a positive attitude perceive basic training as easier, regardless of their branch of service or personality.

What is basic training?

Basic training, or Initial Entry Training (IET), is a strict and demanding military indoctrination course for new enlistees lasting between 9-12 weeks. Training is extremely restrictive. Recruits put in 16-18 hour days and have little personal time. All new military personnel must go through some form of rigorous formal training. The primary purpose of "basic" is to change a soft civilian into a disciplined soldier. Changing a person's civilian habits is difficult and takes drastic measures that are highly uncomfortable for the individual.

Sometimes, basic training is combined with Advance Individual Training (AIT). AIT is the specific training a recruit receives which is relevant to his/her military job. If your basic training and AIT are combined (called One Station Unit Training, OSUT) then your total training time at one base is much longer. Generally speaking, combat-type jobs have shorter training terms than technical jobs. For example, an infantry recruit can complete his/her total training in about 14 weeks. In contrast, a helicopter pilot takes approximately a year of training. The length of training is important to keep in mind because training usually entails much more discipline and time than regular active duty service.

Last, it is important to remember that basic training is not like normal military life. Initial training is meant to be stressful, and tough. After you graduate basic, active duty life starts to resemble a rational life with less pressure. Basic training is a special event lasting only a relatively brief time, so do not think your entire military service will be difficult. In other words, the military starts becoming "fun" after basic training.

A word about the Initial Entry Training (IET) book.

The IET book, or similar training book, is given to recruits at the beginning of their basic training cycle. You only get one book, so guard it like your last canteen of water in the desert. If you lose your assigned IET book, you will hate life during training.

The IET book frankly describes all military tasks you need to complete while in training. Your drill sergeant will make you read and memorize parts of it every day. My advice is to get a copy from your recruiter before you ship out and start memorizing it early. Sometimes recruiters do not have a copy, but you can track one down from a prior service person, the public library, or the internet.

Some readers mistakenly believe my guide is just like the IET book. While some of my information is the same, the IET handbook does not supply a full range of helpful hints or insights like this guide. For example, you won't find little things that will make your life much easier like bringing extra black liquid wax to basic training. Additionally, the IET book will not tell you how to avoid trouble with the drill sergeants, or delineate the psychology behind some military activities. In short, the IET book is immensely valuable. However, my guide should be viewed as a crucial supplement to the IET book. Used together, both books provide a solid means to make basic training easier.

Remember, the two most critical items you need is sound information about basic training and a positive attitude. The more you have of each, the more likely you will succeed. After reading this guide entirely, if you decide the military is right for you, I wish you a sincere GOOD LUCK!

Special Notes: Terms and Disclaimers

For the purpose of clarity and simplicity, this guide utilizes the term "basic training" as being interchangeable with "basic" and "boot camp". Technically speaking, only the Marines have boot camp. Another term which has multiple synonyms is "drill instructor," which is interchangeable with (DI), "drill sergeant," and "sergeant."

An additional side note involves a legal disclaimer—I know this is redundant, but we live in a litigious society. Again, please be advised that this book elaborates only on my opinions. The reader assumes full responsibility for his/her actions and decisions. I do not supply advice as a professional, but rather just to illustrate my ideas regarding my perspective of the military.

The reader should keep in mind that the military and basic training cycles are always changing. Although I honestly try to incorporate the latest military practices within these pages, some changes are impossible to update. For example, the pay scale supplied is continually upgraded every October by the government. Therefore, in 1998 a private made $900.90 per month, while in 2003 it was $1150.80. Uniforms, such as the Army's berets and training techniques also change. However, it appears the major training philosophy and structure has been stable for several years. The point here is for recruits to be aware that my guide may not cover everything "new," but the general concepts are sound.

Finally, there is some concern whether it is a good idea to bring this guide to basic training. My advice is to study this book thoroughly before basic training, *but leave it at home*. Drill sergeants usually do not like these types of books. In fact, people inside the military probably do not want you reading this book. You can share this book with friends, or fellow recruits, but it is not wise to be conspicuous with it. Although I have very high regard for some recruiters and drill instructors, my previous guide has been heavily criticized. Some unscrupulous people (hiding behind fake names) have even tried to discredit this book, but do not believe all the critics and criticisms-- trust your own judgment.

Chapter 1

To Enlist, or Not? A Crucial and Difficult Decision

The following statement may seem obvious, but enlisting in the military should not be taken lightly. Enlisting must be founded upon well thought out reasons and specific goals. After meaningful contemplation, you should be able to reflexively state three reasons why you wish to enlist. Your personal motivations should be explained with the utmost sincerity and with little hesitation. While it is certainly normal to have some doubt regarding enlistment, your doubts should be significantly offset by explicit objectives. If you cannot confidently rattle off your reasons, you should reassess your intentions regarding military service.

The purpose of the previous paragraph is to emphasize the importance of a serious adult decision. The military is a highly regimented, strict and sacrificing lifestyle. Basic training is especially austere and restrictive because it must prepare civilians to live within the military. Despite what you may read, or hear in the popular press, military training is still difficult and unlike anything in the civilian world. It does not matter what type of person you are, you will experience stress and challenges throughout basic. Moreover, you give up many civilian rights once you sign your enlistment contract. For example, unlike military personnel, civilians cannot be "ordered" to do anything by an employer and civilians can easily quit their jobs. Indeed, the military is radically different from life as you know it.

Individuals should never feel compelled to enlist by family members, recruiters, or due to current pressured situations. Your decision to enter military life should be based on a positive foundation. Ideally, new recruits enter the military because they *want to*, not because the feel they *have to* enlist. Again, my intent is not to scare people who are thinking about a military career. However, several hours of honest self-reflection and research is better than spending years in an occupation which is not for you. While most people find fulfillment in the military, a few individuals live with intense frustration because the service life is a poor fit for them. My advice before enlisting is to think very seriously and do the research to discover if the military is right for you. Read this book entirely before enlisting so you can get a general feel for the military lifestyle and make an informed decision.

Military Benefits: The good and bad

With the latter introduction being the strongest caution I can provide without frightening new recruits, there are many positive reasons I can give for enlisting. In fact, I endorse the military as an outstanding short-term or long-term career choice for most people. I hope that, this section will supply the reader with several solid ideas why they should enlist. Although your best response to a recruiter, or drill sergeant is, "I want to enlist because I want to serve my country," they know there are other pragmatic reasons for enlistment. Listed next are some (but not all) of the most popular reasons recruits give for enlisting.

Reasons to enlist:

Respected training and job skills. How many good jobs can you name for high school graduates? Flipping hamburgers at a fast food restaurant, cleaning jobs, and low- paying office work are some of the uninspiring jobs awaiting many youths today. You can spend $20,000-50,000 on a college degree and not be assured of a job when you graduate. On the other hand, you can immediately start a respectable career by enlisting in the military.

Decades ago, America had several industries willing to pay its employees during training. Employers offered sound benefits, some job security, and steady promotions. Unfortunately, the climate of American businesses has drastically changed and few employers offer the type of paid benefits as the military. Moreover, civilian employers are willing to drop employees at the slightest sign of economic weakness. It appears there is no loyalty or trust between civilian employers and employees. In contrast, the occupational outlook and job security is much better in the military than in the private sector.

The Army has about 250 different types of jobs, called Mission Occupational Specialties (MOS). (See the Appendix for a partial list of military jobs.) Nearly any civilian type of job can be found in the military. From firefighter to trumpet player, the military usually has an equivalent civilian counterpart. Most military jobs are combat arms related (soldiers, artillery, tanks), although support type jobs are plentiful.

The two prominent benefits under this heading are "choice" and "paid training." It is highly unlikely that people without any experience can choose the job they want and have it provided to them. Second, not only are you free to choose a military job that is of personal interest, the training for your selected job is free. You will receive quality training *and* be paid a full salary while in service. The combination of career choice and being paid is truly an amazing opportunity rarely seen anywhere in the civilian market.

The following example clearly illustrates the military's benefit. Jack has no experience and only a high school degree. While Jack was a stellar student in high school, making A's and B's, he could not secure a job after graduating. Jack has a strong interest in the medical field; specifically he wants to be an operating room technician. Unfortunately, he does not have the $8,000-10,000 dollars to pay for training at a local community college, so he remains working at a low-paying retail job at the mall. However, Jack could enlist in the military to become a medical technician and get paid over $1,100.00 per month while training. Additionally, his housing, clothing and food are paid for in full. After a few years, he could leave the military to take a similar medical civilian job. In short, the military offers Jack a way to meet his goals without skipping a beat, or drowning in a flood of student loans.

Education Benefits

Pursuing a college degree is certainly a worthwhile endeavor. While most people mistakenly believe four-year universities teach job skills, their primary purpose is to stimulate higher order thinking and the ability to reason effectively. The ability to think critically is certainly necessary to expand one's knowledge about the world and to understand complex concepts. However, a college degree does not necessary translate into discrete job skills. In my observation, vocational schools and community colleges teach specific skills, while universities broaden people's general cognitive abilities. When having distinct job skills is married with a university degree, you have an outstanding chance of attaining occupational success. This is not to say you cannot be "enlightened" without a university degree, but higher education does provide a formal framework to this end. Fortunately, the military offers a means by which you can have the best of both worlds. You can learn a solid employment skill, while also securing a college degree.

Educational benefits have been a consistent enlistment enticement since veterans returned home from World War II. The government realizes the benefits of having highly educated personnel. In the government's view, a smarter military is an effective military. While some progressive civilian companies pay for their employees' higher education, such companies are few.

There are many ways to attain a college degree and have the military pay for it. First, if you are already in college, take out government student loans to pay for your school expenses. After you graduate college with your degree, speak to a recruiter about the military loan repayment program. For every year of military service you give, the military will pay back part of your student loans. For example, when I served a few years ago, the Army paid back about 33% of a person's loans for every year of service. If a soldier had $15,000 in student loans, the military paid back about $5000 per year of service. If the soldier served three years, the student loan was paid off in full (5000 x 3 years = 15,000). Note, due to the frequent changes in educational awards, the examples given should be viewed only as a general idea of the loan repayment program.

If you have not gone to college, but wish to pursue a degree in the future, inquire about the Montgomery GI Bill and the Army College Fund. The latter funds can supply you with *up to* $50,000 for future use. Usually, the highest amount of educational funds is only available if you serve a full standard enlistment of four to six years. However, if you sign up for two years (the minimum enlistment contract) you can be eligible for about $20,000.

The following description is how the system works: While serving in the military, $100 is deducted from your paycheck. These deductions are usually taken out of your check for about a year, but it could be longer. The deducted money is necessary for you to be eligible to receive a much greater amount which the military gives you at a later date (it is like fee). Even if you put in only $1,200 for a year, you are still eligible to receive several thousand dollars for college after you leave the service.

The third way the military can pay for your education is by allowing you to attend college classes while in service. The military can pay up to 75% of your school expenses while you serve. Ask your recruiter specifically about the Defense Activity for Non-Traditional Education Support (DANTES). This distance-learning program provides a means to attain a college degree with minimal disruption to your life. DANTES supplies a wide range of courses from high school to college level. Usually classes are "online," or correspondence type of work. Sometimes, you might be able to attend night classes if your MOS permits. Regardless of the type of course you take, the military usually can pay for it.

Finally, if you are in the Reserves, or National Guard, lucrative benefits similar to the active services exist. In some States, the National Guard will pay a full 100% of school tuition! In my opinion, the National Guard offers exceptional educational assistance.

Large Cash Bonuses

The military offers large cash incentives to tempt many new recruits. If personnel shortages exist in specific fields, then the military will give enlistees "bonus" money for selecting the shortage MOS (job). For example, if there is a personnel shortage in artillery, the Army might offer you a bonus of $10,000 dollars. The highest bonus I heard of was about $12,000.

However, there are caveats associated with bonus money. First, you will not receive all the bonus money promised you in a lump sum. Generally, your bonus money will be portioned out to you in small increments every year. To illustrate, let's say you have a $2,000 bonus because the army needed tank drivers and you enlisted for this MOS. You might receive $1,000 of the bonus the first year and the next $1,000 the following year of service, but you do not get all of the money at once. Additionally, you normally have to enlist for a full term (four, or six years) to be eligible for many bonuses. Last, there is a distinct reason why the military attempts to lure recruits into hard-to-fill jobs—the jobs are usually difficult. Such jobs entail long hours and dangerous work. Always ask your recruiter for all the details about bonuses and what strings are attached to receiving them.

Medical and Retirement Benefits

Although people do not enlist solely for the military's medical or retirement benefits, these two areas are exceedingly significant perks the military offers. With health care costs approaching almost $600 per month for a family of three, the military offers people a sound way to provide health care for their family. More civilian businesses are passing the cost of healthcare to their employees, but the military is not. The level of professional care provided by military hospitals has impressed me. Additionally, many doctors are trained in highly competitive American medical schools. While it is true patients might have to wait for medical service slightly longer than civilians, it is largely free for soldiers.

Another big benefit concerns retirement. It seems secure retirement is non-existent in the civilian world. However, in the military, a person who serves for 20 years can retire with a stable paycheck every month. For example, if you serve 20 years, you retire with 40% of your pay. This percentage goes up for every year past 20 years. In addition to a consistent paycheck, the retiree retains medical coverage. Smart people who enter the military at 18 and serve a full 20 years can retire at only 38 years old!

Confidence, Leadership and Responsibility

The military expertly trains leaders. It is amazing to watch a 20-year-old person command a squad of men/women and the squad obeys every command given. Leadership and mature responsibility are intangible assets which can last a lifetime. If you learn solid leadership skills, chances are you can capitalize on these important skills within the civilian world. People who have been instilled with the "can do" military spirit usually become successful in their future careers. Do not underestimate the immense power of confidence the military provides its recruits.

World Travel and Excitement

Many recruits enlist because their life of fighting imaginary monsters on their Sony Playstation is boring. The military offers experiences the civilian world cannot. For example, the military is the only place you are sanctioned to plow over trees in a 65 ton Abrams tank, use C-4 plastic explosive to blow up bridges, and fire assault weapons. If you want excitement and adventure, choose the appropriate combat MOS and you will surely find it.

Additionally, if you enlist for more than four years, you will most likely be stationed outside the United States for part of your tour of duty. Opportunities for travel are excellent in the military. My younger brother, who was a combat medic, was stationed in Korea and other parts of the world. His experiences overseas broadened his perspective and enriched his life. Traveling is truly the best type of education a person can receive.

Increased Standard of Living

Do not believe news stories about how military personnel struggle financially, or how many soldiers are on government assistance. The truth is, only a small percentage of military personnel receive assistance because they have foolish spending habits. For example, it is not wise to have six children and two ex-wives on a buck sergeant's salary.

If you manage your money wisely, the military offers the opportunity to live a good life—especially for single personnel. (See pay schedule in the Appendix) To be certain, you will not get rich, but you will acquire job security and decent money. The military provides for big expenses such as food, clothing and housing. If you do not have to pay for insurance, food, or shelter, then you can save this money for other items. The lowest ranking person (Private E-1) during basic training earns roughly $1,070 per month (year 2003).

It is an unfortunate fact of life in this country that many new recruits come from impoverished conditions. Living in run-down trailers and trading food stamps is not a good life. Indeed, the military offers a highly respectable means of leaving undesirable conditions.

Physical Conditioning and Challenge

I put this benefit toward the end because I personally believe there are better reasons to enlist than getting into good physical shape. Certainly, basic training will make lazy individuals stronger and more capable, but it will come at a hefty price. Some people who enter the military to get healthy and strong realize they cannot leave once they have attained this goal.

Other people want to prove they have what it takes to make it through basic training. I find it interesting that many people who have thought about the military, but did not join, later regret it. Some individuals who had the chance to serve, but did not, are subtly bothered by chronic unspoken regret. Such people ask themselves, "Could I have met the challenge of basic training? Was I tough enough?" I know these latter statements sound hokey, but I have friends who are troubled by the fact they will never know if they had what it takes.

Duty, Country, Honor, Patriotism

The one reason recruiters and drill sergeants love to hear as to why recruits enlisted is for patriotic reasons. Most recruits do not sign up for this reason, but patriotism becomes a reason during your military service. While many people talk about the words duty, honor, and country, recruits act on them. Patriotic terms take on a much more significant meaning when you proudly wear a United States military uniform. Military patriotism is something I cannot describe; it must be experienced to fully convey its meaning.

Reasons Not to Enlist and Other Considerations

For all of the previously stated reasons to enlist, remember the venerable adage that nothing worthwhile in life is free. While the military is a superb means to achieve various goals in your life, the government expects much from you. Military jobs, while similar to civilian jobs, have additional responsibilities and restrictions. As I stated earlier, the military is not for everyone. Listed next are some reasons you should reflect on and ask yourself if you can live with them.

Restrictions and Sacrifice

I initiate this section with what I believe is the most important consideration regarding military life. Most recruits do not fully grasp what the term "strict" really means. In the military, you are *told* what to do, not asked. Especially during basic training, you are told when to wake up, what to eat and what to do. If you cannot take orders well, do not join the service.

Many times, recruits will be told to do things they do not like, but they *must* obey. Even after basic training, you are always under the strict supervision of a higher ranking person. You can be the most intelligent and most athletic person in your entire platoon, but still you must obey orders from someone with the IQ of a houseplant if that person has a higher rank than you. To be fair, chances are that your platoon leader and company commander are well educated and will usually treat you with respect. Also, you cannot be ordered to do something which is illegal. Still, most people will find themselves being told to complete highly undesirable tasks, or suffer consequences worse than merely being fired from a job.

The military restricts your freedoms not only by its rank and order system; it enforces military rules via severe consequences. Let us say you are late for morning assembly two times this past month. Although you may have only been one minute late to formation, your drill sergeant thinks the proper punishment is for you to clean all the latrines (bathrooms) in your barracks. Additionally, your sergeant orders you to clean toilets for an entire month. If you refuse to clean the toilets for the assigned month, you can be given even more harsh duties and a reduction of pay. You may even be ordered to work night shifts, or be confined to quarters. Unlike a civilian job where one would simply tell the boss he/she is crazy and quit the job, you cannot quit the military. If you attempt to leave the military without authorization, or disobey orders, you can go to military prison!

There are many military rules which are in stark contrast to the civilian world. For example, you must wear your uniform and hair a specific way. You must meet very strict physical fitness standards. There are things you cannot state publicly, like speaking out against the president. Soldiers must obey military protocols like saluting and standing at attention. You must be ready for surprise inspections at your barracks. You can be ordered not to go to certain places, even on your free time. The bottom line is that once you enlist, you do not have the same rights as a civilian. All military personnel are governed by the Uniform Code of Military Justice (UCMJ). Although military law is different than civilian law, soldiers can be tried for violations of the law in both civil and military courts. The danger with being tried in two court systems is that a person can actually be punished twice for one infraction.

The sacrifice portion under this heading also entails being separated from friends and family. Many times, your military orders assign you to a place where you cannot see loved ones. You may find yourself on two week-long assignments in remote areas which cannot be accessed by non-military personnel. Worse, you may be assigned a "hard-ship" tour of duty where you are separated for six months or more from friends and family.

After initial entry training is complete and you are an active duty soldier, you still may find yourself engaged in constant training. Sometimes training exercises can have you working 12-18 hours a day. You may go without bathing, or without a hot meal for days. Most military personnel, at some point in their service, go through a period where they will sacrifice a large part of their personal time.

Of course the ultimate sacrifice military service people make is putting their life at risk. Recruits must understand that they are entering an occupation which could entail them killing another person, or being killed. I knew a few recruits in my training cycle who realized too late that the Army was not a big game and that they were actually being trained to take another person's life. When you are constantly running full charge with an eight-inch bayonet at a life-like figure yelling, "Kill! Kill! Kill! With Blue Cold Steel!", you understand that the military is deadly serious. Even Reservists must be fully aware that they can be put in harm's way within a moment's notice.

Time

Depending on your MOS (job) in the military, you could work a regular eight-hour day. In fact, some military jobs are difficult to distinguish from civilian jobs if it were not for the uniforms and the occasional saluting. However, several jobs require that you report to morning formation before 6:00 a.m. (0600hrs) so you can do morning exercises. Additionally, you could find yourself constantly working weekends and 10-hour days all week.

Although many recruiters tout the educational benefits of the military, you might not have the time to take advantage of some of them due to time restrictions. Some combat-related jobs, such as infantry, have schedules which are not well suited for enlistees to take advantage of all the educational benefits.

Another important time consideration involves vacation time. Depending on your MOS, you may not have as much free vacation time as you think. First, you must request vacation time well in advance and your request must be *approved* despite the fact you may have time credit. Second, while the military sells recruits on "30 days of vacation," many of my friends still on active duty say it is difficult to schedule all of those free days. Recruiters will not tell you about the difficulties associated with the "30 days off." However, several factors go into your work schedule and vacation time such as the type of commander you have, your specific MOS and where you are stationed.

Physical Training Exercise (PT)

This factor could be a reason to enlist, or not to enlist. Obviously, staying healthy and being paid to do so is a benefit. However, some enlistees do not like waking up at 5:00 am every day and being told to exercise, or exercise with a group. You cannot simply sleep in late one day because you are tired. You must report to morning PT, like it or not. I used to be an avid runner, but even I did not like to be told when to run, or to report to a specific running group. Interestingly, research studies have shown that people who enjoy doing activities show a decrease in their enjoyment for those activities once they are told to do them. Again, the amount of PT (after basic) depends on various factors.

Permanent Jobs and Location

This heading specifically addresses two common complaints. First, you cannot change jobs if you do not like your MOS. For example, if you enlisted to become a tank driver, but found the working conditions too noisy, you cannot change your

MOS. You might truly hate your current MOS, but you must stay with it until your contract expires. Again, quitting, even for good reasons, is usually not allowed.

Second, the military places you where it needs you. Although you might want to be stationed in Europe, you could find yourself at Ft. Riley, Kansas for three years. Additionally, you must live within close proximity to the base you are assigned. Base living is sometimes hard on families because they are not in desirable locations. Accommodations, while adequate, cannot be described as luxurious. If you are single, you could find yourself stationed in a rural area and rooming with three other people you do not particularly like. You can always request a transfer to a different location or barracks, but personal preferences take a back seat to the military's priority.

It is imperative that the MOS for which you enlist is the job your truly want. Select your MOS very carefully and not just because it sounds "cool." If you wish to be stationed at a specific location, this can sometimes be arranged. You can have a station of choice written into your contract. To reiterate, make every decision carefully and well informed.

Frequent Moves

Let us say that you love your first duty station at Ft. Hood, Texas. You have great friends, you met your spouse in a nearby suburb and you hang out at exciting places on the weekends. For the last three years you are thinking this is the best place to live for the next 20 years. Everything is great until one day you get your "orders" to move to Kentucky. Your spouse tells you she cannot leave Texas, your kids like their school and you do not want to live in the sweltering humidity of Kentucky. Unfortunately, many military families are uprooted suddenly and ordered to move their lives somewhere else that they may not like at all. Even some single personnel have a hard time adjusting to moving every few years. While traveling is an incredible experience, it is nice to call one place home.

Hiding Out

If you are tying to hide from the law, or from prior obligations, do not think you can hide by enlisting in the military. While many people think they can seek refuge in the massive ranks of the armed forces, their past usually catches them quickly. I remember a person being taken out of the processing station because a records check revealed the person lied about his past. The person in question had a warrant out for his arrest due to a misdemeanor drug possession offense. The military will perform a formal records check to see if you have been arrested, or have outstanding violations. Even if you have just a speeding ticket on your record, tell your recruiter everything before enlisting.

Another example regards child support responsibility. Even though this is anecdotal, I heard about a soldier who tried to get out of paying child support by enlisting and changing his name. Needless to say, this soldier was separated from basic training faster than you can read this sentence. The bottom line is to be honest with your recruiter. Just because you may have a rough history, it does not necessarily mean you cannot enlist. However, if you lie and conceal your past, you are headed for trouble.

Help With Making the Big Decision

In the rectangle box below, write your primary short or long-term *life* goal (s). Next, within the square box, write three reasons why you want to enlist in the military. In the opposite box, list three concerns you have about military life. After the task is complete, think hard if your reasons for enlisting outweigh your concerns. Ask yourself sincerely if you can live with your concerns. Determine if your reasons to enlist will get you your major goal. To be sure, there will be some doubt in your mind, but remember, reaching a worthwhile goal takes effort and some degree of discomfort. The real question now becomes, how much discomfort are you willing to endure for your goals? The more valued your objective, the more discomfort you should endure to reach that objective.

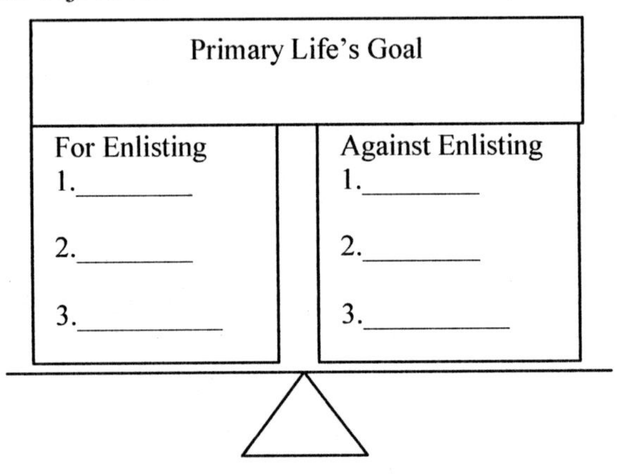

Still Having Trouble Deciding?

As mentioned, joining the armed forces is a significant decision in one's life, so do not worry if you are finding this process difficult. If you are still having difficulty, talk with a career counselor at your local public high school or college campus. This is not superficial advice. Career counselors are exceptionally trained to assist people making vocational decisions. Usually, you will find guidance counselors extremely friendly and eager to help you. However, make sure your counselor is certified, has a few years experience and has a college degree in counseling.

I would also suggest you take a career assessment and personality test. Standardized assessments will ask you critical questions which can illuminate your true interests and inclinations. There are many outstanding vocational tests on the market, but choose wisely. I can suggest the Kuder, or Strong Interest Inventory. Also, the Self Directed Search (SDS) is a good test choice. Valid personality tests are the Catell 16PF and the NEO-PR-I.

Generally, career counselors will help you with any formalized assessments or testing.

Last, seek out people who have been in the military within the last five years. Ask neighbors, older relatives or family friends if they know of anyone who served in the military. If you find an active duty member, ask him/her if you can interview him/her for about an hour. Your interview can be simple, but have several questions written out beforehand. Your question can even be, "What are a few things you liked and disliked about the military?" Additional sources of information are favorite teachers, or coaches. You might even try asking a recruiter to put you in contact with someone recently graduated from training.

Being Decisive and Finalizing It

Assuming you have sincerely put a few weeks of effort and deep thought into your military research, you now have tangible facts before you. A few people might think you are crazy for joining, while others support you. However, you should have other more objective sources such as your vocational guidance test, and your goals which point you in a good direction.

The key to your mental health is to make a firm decision. Your decision to enlist or not to enlist in the armed forces must be final and conclusive. Do not be flimsy in your judgment; it will only cause you grief and regret.

While it is acceptable to delay a major decision for a legitimate reason, make sure your delay is not an excuse. For example, you may delay enlistment due to a family member's illness, or to finish school. Simply put, if your reasons for joining are sound, make a specific date to enlist. If you are nagged by the concerns you wrote and you are full of doubt, do not enlist.

If you decide *not* to enlist, but are still interested in the military, tell yourself you will not think about this issue for a full year. Set a concrete date after a sizeable amount of time to revisit the idea of military life and to re-read this book.

The benefit of being definite and resolute is that it staves off serious anxiety and remorse. Whatever your decision, do not look back and dwell on it. Go forward, develop a positive attitude and do what is necessary to maintain your mental well-being.

Chapter 2

First Step: The Enlistment Process

This chapter assumes you have made a well-informed decision and decided to take the challenge of military life. You feel confident that the military's benefits can help you reach many of your stated life's goals. Rest assured, with the right attitude you will find excitement and adventure in the next few years. Indeed, many people in the military speak fondly of their robust experiences for the rest of their lives. Basic training, especially, is an event no one can ever forget.

This chapter is designed to take a prospective recruit through the enlistment process. You will meet with recruiters, take an aptitude test, take a physical examination, and come into contact with the military structure. You will notice most of the people within the enlistment process are wearing uniforms, are more serious than usual, and follow a strict sequence to process you. At first, the military atmosphere seems strange and slightly frightening. However, different work environments often are a little anxiety provoking. My advice during this stage is to relax, and realize that some anxiety is normal. If your intentions are planned and you have an objective, the first step in the enlistment process will go smoothly for you.

The Recruiter

Over the years, I have had contact with several recruiters. Some recruiters are better than others in their advice and level of selling pressure. On average, most of today's recruiters are honest and very hard working individuals. Recruiters are usually carefully selected people based on their relevant skills and appearance. They are also highly trained and educated professionals bent on discovering what makes a prospective recruit interested in the military. Always keep in mind the primary mission of a recruiter is to make you join the military.

The days of recruiters' "hard sell" and shady practices are largely extinct. Now, the government spends millions of dollars every year on flashy commercials and a variety of enticements to lure people into the military. From the moment you walk into a recruiter's office, everything in the room is specifically designed to "sell" the military to you in a softer unconscious way. For example, the highly polished uniform of the recruiter, the action-oriented posters on the wall and a banner touting $50,000 for college are all designed to make a subtle impression on you. From a psychological point of view, the polished image of the recruiter reflects that this person has his/her life together, the action posters imply adventure or excitement, and the money for college provides an avenue to attain a highly regarded social goal.

Every year, the military *needs* tens of thousands of new people to join. In recent years, the armed forces have struggled to meet recruiting goals—despite what you might hear to the contrary. Therefore, recruiters are commonly under pressure to find and motivate prospects to enlist. Again, the majority of recruiters are decent people regardless of the persistence of old Hollywood stereotypes. Recruiters are not liars and will not use fraudulent means to make you enlist. However, it has been my experience that recruiters only sell the positive aspects of military life. Recruiters certainly downplay the negative parts of service or will attempt to reframe an undesirable part into something good. Most important, it appears to me that recruiters will not elaborate on information if not asked. In other words, you have to ask pointed questions, or the recruiters will continue to show you only the brighter side of the military.

Recruiting Techniques

Many of the recruiting techniques are similar to those found in the civilian world. Go to a shopping mall, or analyze TV commercials and you can see the subtle ways businesses try to sell consumers goods or services. The following techniques are not intended to dissuade you from enlisting, or to anger recruiters, but rather to make the reader more aware and better informed.

(1) Challenge, image and group identity: A recruiter may portray the military as an elite occupation only worthy of a select few types of people. "You may not have what it takes because the Marines are only for the few and the proud. Marines are extremely tough and not many can wear the uniform." Admittedly, this previous recruiting quote does not work with everyone, but it does evoke emotions in most. The purpose of this idea is to challenge the person's concept of self-worth and self-image. Recruits may think if they accept the challenge, then they are part of the worthy, tough group. Fear of being "common" is a significant motivator to join a group which is not common. The need for affiliation with a "special" group is a basic human need. Recruiters sometimes try to challenge a person to think about their current group identity and to compare it to the polished warrior-like image of the military.

(2) Emotional Excitement: "I bet you would love to repel out of a Blackhawk helicopter!" "How about screaming across muddy fields in a massive 65-ton tank and blasting away trees with its 120mm cannon!" "Did you love the movie *Top Gun*? Only the hottest babes really dig guys who fly fighter jets." These previous pitches are all designed to tap into your emotional centers. To be sure, every branch of the armed services can offer you some excitement. However, a favorite selling technique used by the automobile industry is also used by recruiters. Once a salesman gets you emotionally stirred regarding a new vehicle, your ability to think objectively declines. You may then start to deviate from your closely held budget to purchase the new car because you are not thinking rationally.

Only recently have neuroscientists uncovered the biology behind how emotions affect the logical centers of the brain. Without delving into too much detail, research has demonstrated that when a person becomes emotional (excited, angered, or depressed) the emotional parts of their brain actually override the logical thinking centers.

Usually, if a recruit says he is interested in a combat-related MOS, the recruiter has a very stimulating high technology video to show him. The purpose of the video is to excite you and to get you to think emotionally, not logically. To be honest, nearly everything you purchase has some degree of hype. This is not to say excitement and adventure are not important factors, but try to balance this aspect with other meaningful goals. Remember the emotional technique and think with a clear head when speaking with a recruiter.

(3) No other choice: Once a recruiter has a grasp of your personal history, he/she may paint your current situation in such a light that you feel the military is your only choice. "Look, you only have a high school degree, you did not make great grades, you have no experience or skills and you cannot get a decent paying job. Your options are few, but the military can get you out of your current situation and into a respected life." Although, the recruiter in this quote might be right in that the military offers a good avenue out of an undesirable life, there are usually other ways to gain success. Rarely is there just one solution to a problem. I do not want to add to the doubts of prospective recruits, but do not be forced into believing you only have one option, or that your life will not get better if you do not enlist.

(4) Money and other enticements: While offering money to enlist may not seem like a sophisticated technique, it has a very strong appeal. As mentioned before, the military offers good money for education and for jobs. However, the price for securing a steady paycheck or money for school is significant. You will work hard and have many personal sacrifices for the military's monetary benefits.

(5) It is your duty: "Only true patriotic people who love their country serve in the military. If you do not serve in the military, you really do not love your country!" If a recruiter tells you the previous statement, do not believe it. It is unfortunate that many civilians think that a person must serve in the armed forces to be a true American leader, especially in political life. Military veterans should always be shown the highest respect. However, loyal citizens who have not served should also be shown respect. I have met exceptional people within the military, as well as civilians with no military experience. The true measure of a people are their character and deeds, not their job.

Your First Meeting with a Recruiter

Even if you have not made a firm decision regarding enlistment, you should speak with a military recruiter to gather information. Despite the "selling" nature of recruiters, the information they provide will be true and current. The easiest way to locate a recruiter is usually the phone book. Recruiters are located under various sections in the phone book, but search the blue government pages first. Typical headings include, "Armed Forces," "Military," "United States Army (Marines, Air Force, Navy)," "United States Recruiting," and "Recruiting." If you are searching the Internet, go to the Internet address, www.goarmy.com, or www. go (branch of service).com. Currently, I use a search engine called Google. If you go to Google.com and type in the keywords, "Military Recruiting," you should get a list of relevant sites. You could also type in the branch of the service you like such as "United States Marines."

Recruiting stations are located in every major city (such as Denver, Houston, Los Angeles), and most large towns (such as Lafayette, Lakewood, and Denton). In other words, it should not be a major issue finding a recruiting office which is within 15 miles of your house. Once you have located a nearby office, call a recruiter and ask to set up an appointment for a personal visit and interview. A good time to schedule a visit is around 10:00 a.m. A mid-morning interview is good because you are not still waking up and you will have the rest of the day to think about your conference.

The night before you speak with a military recruiter, write out specific concerns you may have about the military. Bring your three personal reasons to enlist and not to enlist to the interview. Ask the recruiter for specific details regarding the military benefits which appeal to you. I strongly urge you to write down your questions beforehand. Do not go empty handed. As I mentioned, the recruiter may not volunteer all information which is important to your situation. You must direct your interview by asking good questions.

There is no need to be excessively nervous or defensive when interacting with a recruiter. Recruiters are professionals and have seen hundreds of prospective enlistees. Recruiters will answer all of your questions truthfully—so do not hesitate to ask questions which could impact your future life. This is not a time to be timid. The interview will go smoothly if you are sincere and have a semi-structured format to follow. I advise people to think of the recruiting station visit as a fact-finding, or informational interview. This is a time when both parties can see if the military is a good fit for you.

You might think the following does not need to be explicitly stated, but be aware of your appearance before the interview. I have seen many people enter a recruiting station dressed improperly. Do not overdress or wear formal attire, but you should be well-groomed and presentable. Do not wear torn jeans, shorts, or tee-shirts. People will likely treat you with respect if you look respectable.

Upon entering the recruiting station, which looks like a nice business office, your recruiter will be very friendly. He/she will shake your hand and offer you a beverage. Your recruiter will be a sergeant (E-5, or E-6), not an officer. You will sit at his/her desk which is usually part of an "open" room. As you sit down, notice the posters, the uniforms and the atmosphere of the room—it has a purposeful design. Your interview should only take about 30-60 minutes, so start asking your written questions early. Make sure you write the sergeant's responses down accurately. Your questions should give the recruiter an idea of what interests you have in the military. Of course, skilled recruiters are attempting to figure your motivations out even before you sit down. If you say you are interested in money for college, rest assured the recruiter will have a nice presentation to show you about educational benefits. Interested in adventure? He/she will show you an action-packed 10-minute video. Whatever personal goal you reveal, the recruiter will have a package to show you.

After the interview is over, make sure you shake your recruiter's hand and thank him/her. Again, be pleasant and professional. Upon leaving the recruiting station, go someplace to relax and review your notes. Reflect on new information you did not know before your interview. Ask yourself about additional concerns which may have been brought out. Did the recruiter address all your questions effectively? Do you need to ask any follow-up questions? If, after your interview with the recruiter, you are still interested, call your recruiter again to ask any unanswered questions and to schedule an appointment to take the ASVAB test.

The Second Interview and the ASVAB Test

During your first informational interview with a recruiter, he/she will ask you for basic information such as your age, phone number, and the like. Usually, a recruiter will call you the next day or two to ask you if you are still interested in enlisting. If you are interested, you will take a free test called the Armed Services Vocational Aptitude Battery (ASVAB). To better understand your interests, and to determine which military jobs you qualify for, you must take the ASVAB. All prospective recruits take this test at the recruiting station, or at a government building called MEPS.

The ASVAB is a series of paper and pencil tests which take approximately three hours to complete. This long test is administered under very tight supervision and cheaters are quickly removed. You will receive a few breaks during the test, therefore do not panic about the length.

The ASVAB is a widely accepted and well-validated assessment. This test has questions regarding electronics, mechanics, math and other subjects. Some tasks are timed to gauge your speed with processing information. After you finish your test, a computer will score your responses and provide results the same day, or very shortly thereafter. Your recruiter will explain your scores in detail to you.

Insider tip: Even though taking the ASVAB in no way obligates you to enlist, the military environment in which you take the test makes you think differently. Do your best to relax and focus on just doing well on the test. Never leave any item blank—guess on items you do not know. When the test administrator informs the class there is only one minute left for a given sub-test, start filling in all the blank ovals with your *best* guess. I strongly advise you to prepare for the ASVAB because the better your score on the test, the more military jobs you can select. Your public library and local bookstores (Barnes and Noble, Amazon.com) have several books specifically written on the ASVAB. I suggest buying Baron's, or Arco's study guides. One of the most effective ways to study for the ASVAB is to take as many practice tests as possible—these are available in the study guides mentioned. Some recruiting stations even have a short practice version of the ASVAB you can take which predicts your full-scale score. In addition to the academic preparation, make sure you eat a solid meal beforehand and definitely get at least eight hours of sleep the previous night. Also, drink a moderate amount of water (not soda) immediately before taking the test.

Although ASVAB's are administered differently around the country, the following is an example of how it usually works. Your recruiter brings you to a government building called MEPS (Military Entrance Processing Station) to take the ASVAB. MEPS is a busy place with sergeants scrambling to process new recruits into various branches of the military. You arrive at MEPS between 5:30-6:00 a.m. to take the test. You may eat a free breakfast first, but afterwards you are escorted to a large classroom with other prospective recruits. Some recruits are entering the Marines, and others the Navy or Air Force. As you sit in your desk waiting, an active duty soldier will instruct you as to the seriousness of the test and the penalties for cheating. Keep in mind that MEPS is a secured government facility and rules are strict. After reading word for word the ASVAB's instructions, the proctor hands out the test booklets to the test takers. Listen carefully when the proctor says "begin" and "stop." Actually, you will not have any trouble hearing the proctor because he/she usually yells the instructions. Make sure you follow exactly what is said. Midway through the test, you will be given a short 10-15 minute break. After the test is complete, you might be served a free lunch. After eating, your recruiter will pick you up from MEPS--again MEPS is a very secured facility and you cannot just walk in and out freely.

Your recruiter will go over every detail of your test results with you. He/she will tell you which MOS fields are available to you based on your ASVAB results. Typically, most people can select several jobs, about 75% of the jobs available. If you scored low and do not meet the minimum requirements for a MOS of your choice, take the ASVAB again. You are allowed to re-take the ASVAB if you are not satisfied with your performance.

Never settle for a second choice MOS job selection unless you truly want it. If you have genuine multiple interests, a second option MOS is acceptable for you to take. However, never be "talked" into a job which does not feel right for you. Also, never be lured into an MOS just because of the large cash bonus it offers. Either retake the ASVAB to secure your favorite MOS, or rethink your entire decision to enlist.

Your Physical Examination and Enlistment Contract

This is the final stage in the enlistment process. You have thought out your decision, clearly stated goals, gathered facts from your recruiter and have taken the ASVAB. You are now ready to visit MEPS a second time to take a full medical physical and to sign your enlistment contract. (Sometimes the ASVAB and your medical are given in the same day, which makes for a very long day). You should not have nagging doubts at this point. Do not enter MEPS a second time if you still have doubts about joining the military for the next few years.

As with the ASVAB, you will take your medical examination with a large group of people. The process is strictly controlled and this is the first time you are exposed to military structure and order. There is a lot of "hurry up and wait." You might become a little nervous as you are ordered (not asked) to take off your clothes and walk from room to examination room. I advise you to wear "normal" underwear during your physical because for a long time you will be poked and prodded while undressed.

To screen for diseases and drugs, you will have blood taken from you and asked to urinate in a cup. You cannot foil the military's drug screening, so do not try. You will have your hearing checked by putting on earphones and being asked to press a button every time you hear a certain tone. Your eyes are examined, as well as your muscle tone and weight. A doctor or physician's assistant will ask about your medical history, previous illnesses, and drug use. The most humorous part of the physical is when new recruits are asked to walk like a duck in their underwear around a large room. I think this latter exercise tests for leg problems, or other movement disorders. Finally, to put nervous recruits at ease, the doctor does not stick his fingers up your rectum. Sure, the doctor will have to look at your nether regions, but there is hardly any touching. In the end, your medical physical lasts about two-three hours; it is thorough and well- documented in your personal records.

About 30-45 minutes after you complete your physical at MEPS, you will meet with a military guidance counselor. The counselor, usually a mid-ranking sergeant, will have your recruiter's records when you meet in his/her office. The counselor will know your interests and what type of military jobs are available for you. He/she will ask you a pointed question, "What type of job do you want in the military?" The counselor will also ask when you would like to ship out for training (spring and early fall are the best times for training). It is a good idea to have at least six weeks before leaving for basic training. Ask about the Delayed Entry Program (DEP) which gives you more time to prepare for training.

Given your responses to the counselor's questions, he/she will check if you qualify and then turn to a computer to see what job "slots" are available. For example, if you decided you want to be a combat medic, the counselor might say the next available training slot does not start until two months. If you approve the time frame for training and the type of job, all that is needed is your signature on the enlistment contract.

The enlistment counselor will answer all remaining questions you have about training, pay and the details of your new MOS. The counselor will especially explain the details of your contract which you will sign. This is the most crucial step in the enlistment process. Once you sign the contract, there is no turning back. The contract is a legal and binding document. Your contract *obligates* you to everything spelled out in the document. Similarly, the enlistment contract also obligates the military to live up to its promises. <u>The significance of this previous point is that whatever is promised to you must be put in writing.</u> I have underlined this last statement because it is the foundation of your military life.

For example, a recruiter may tell you the military will provide $20,000 for your education, but if it is not in your contract, you will not receive it. A sergeant may promise you that you will be stationed in Europe, but unless the promise is in writing, forget about it. A true example directly affected a friend of mine while in basic training. My fellow recruit's enlistment contract clearly stated that he had to be released from training on a specific date. He was a reservist and was going to school while serving part time in the Army. While the rest of his platoon was slogging it through the final stages of basic, my friend was allowed to leave early and relax. His early departure was fortunate because the drill sergeants did not like him. In any event, the major idea is to have any promise or benefit clearly spelled out in your enlistment contract.

Your meeting with your enlistment counselor should last approximately 30 minutes. After signing your legal document, you will receive a copy. I suggest you cling to your contract like it was stapled to your hand. Also, make a copy of your contract and give it to a close family member for safekeeping. Always make copies of your military documents and keep them in a safe place.

Once all the paperwork is completed, you will be ushered into a room with other recruits for the official oath. Usually an officer will ask you to stand, raise your right hand and repeat after him/her. Then, the officer may casually speak with each recruit, or make a few remarks about the virtues of the armed forces. Once you make your pledge, you are officially sworn in as a new recruit.

Your recruiter will be beaming and smiling when s/he takes you home from MEPS. Your recruiter may even take you out to eat at a fast food restaurant after leaving the processing station. However, while your recruiter reiterates what a great decision you made for enlisting, you feel like someone has tied your shoulder muscles in a knot and that lump in your throat is the size of a cantaloupe. Your mind races with the thought of, "My god, what have I done! I've just signed up for the military!" My advice is to relax. Many recruits feel some degree of "enlistee remorse," but this emotion will pass in a few days. A big change in life is normally associated with nervousness and apprehension. You have made the decision to enlist and you followed through with your decision. To ease your mind, remember that thousands of people go through basic training every month and do just fine. You will also be okay because basic training is hard, but certainly not impossible.

What Next?

After you have completed the long journey from thinking about enlisting, to actually signing your enlistment contract, you will usually go back home for a few weeks. It is not uncommon for new recruits to start training a few months after enlisting. The time span you have between enlisting and "shipping out" for training is absolutely vital. This interim time period is not free time; you must use it wisely as the next chapter demonstrates. Also, many new recruits visit their recruiter regularly for pre-basic training advice and physical conditioning tips. Your recruiter will tell you when you should receive your official "orders" to ship out.

The day you ship out, or the day before, you will again visit MEPS for a spot physical and drug screen. Then, you will load into a bus/van which will bring you to an airport, bus terminal, or rail yard. From either area of transportation, you will be officially flown, or driven (at government expense) to your basic training base.

It is extremely important that you are not late for departure. Sometimes, your recruiter, if he/she is adept, will keep you posted about important dates. However, it is your responsibility to know what you should do. The penalty for being late for training is very unpleasant. Moreover, if you change your mind and try to skip out on your training obligation, legal action will ensue.

Additional Suggestions

If a person asked me whether they should enlist in the military, my first response to that person would be, "Give me three good personal reasons you should enlist." I would also inquire about a person's long-term goals and the level of emotion placed upon those goals. Does the person wish to make the military a career? Does the person want to be a civilian professional in the future? What are the person's educational aspirations?

Unfortunately, many young people do not engage in deep self-reflection and do not know the answers to the few questions in the above example. In my opinion, some individuals want to serve in the armed forces, but are not sure they want to stay in the service for many years. It is not unusual in today's society for people to explore several career paths over their lifetime. Because many people are not sure about the occupations they want, I would recommend that a recruit enlist for the shortest time possible. (Recruiters probably will not appreciate me saying this.) In the Army, this is two years; in the other branches it is about three to four years.

My rationale for recommending a short-term enlistment takes into account the strict obligations associated with the military. If a person makes an honest mistake and finds military life not a good fit for him/her, then he/she can leave after only two short years instead of the standard four years. A short-term enlistment acts as a good "sample" of the military. After all, you do not buy an automobile without first taking it for a short ride. Second, even though a shorter term does not garner all of the benefits/bonuses of a full-term contract, you still receive a large portion of them. Additionally, short-term enlistments sometimes keep you within the United States. If you do not wish to be more than a few hours by plane from family and friends, being stateside is desirable. Last, if you enjoy the military after a short-term enlistment, you can usually re-enlist. Most re-enlistments entail bonuses and therefore you can recoup any bonus money you forfeited by not signing a standard term contract.

Another recommendation I make frequently to people is to explore the opportunities in the Reserves, or National Guard (both branches are sometimes considered Reserves). A recruit in the Reserves/Guard is actually an active duty soldier while in basic training. There is not a difference in the training Reservists/Guardsmen receive than their active duty counterparts. In fact, sometimes Reservists train side by side and you will not know the difference between who enlisted to be active duty and a Reservist. After basic and advance training is completed, the Reservist goes back home, while the active duty soldier is stationed at a military base for his/her full time military job.

Reservists/Guardsmen are part-time military personnel. After active duty training, they usually report to a base only one weekend per month. In the summer, Reservists have a two-week training event. Reservists drill and maintain proficiency in their MOS during their weekend training. They are paid and receive partial benefits. The biggest benefits are usually education-related—some National Guard units pay 100% of school tuition. Even though there are limitations, the Reserves/Guard is the least restrictive way to experience the military. Moreover, a Reservist/Guardsman can *sometimes* enter the active duty ranks if they like their experiences.

People entertaining the idea of Reserve service should note that not all military jobs are available to them. For example, if you have an interest in military intelligence, but your local unit is an armor unit, you may be out of luck. However, it is not uncommon for Reservists to drive to the next city/town to a unit which has a job they enjoy. Also, note that most National Guard units are combat-arms-related, while the Reserves are usually support-related. It is believed by many that the Reserves are called up to active duty more than the National Guard. With either branch of service, people should be fully cognizant that they can be called for duty at any time and you *must* leave your civilian job when ordered.

My last supplemental point regards your MOS selection. My military job was infantry (11 Bravo), which was exciting and full of action-oriented tasks. However, if you plan on re-entering the civilian world to apply your military training, you might want to think about enlisting for non-combat specialties. After all, what can an artillery crewmember do in the civilian world? Combat jobs are sometimes fun, but the skills you learn in combat arms only indirectly apply to civilian jobs. Police departments, SWAT teams, and other law enforcement agencies are the few career choices if you have a combat MOS.

Summary

The following section is a recap of the major points to keep in mind during the enlistment process. Although the summary might be redundant, it is provided to emphasize points which can greatly impact your life.

1. Think deeply about your goals and how the military can meet them. Be specific about your objectives and write them down.
2. Your initial visit with a recruiter should be viewed as an "informational interview." Your recruiter is a professional, but his/her job is to portray the military in a very positive light. You must ask the proper questions which are relevant to your life.
3. Be aware of the selling, or subtle promotional techniques used by the military. Think logically, not emotionally.
4. Do not lie about your personal history. Be up front and honest with your recruiter. Infractions of the law will not necessarily bar you from enlisting.
5. Study for your ASVAB test by taking multiple practice tests available in Baron's or Arco's study guides. Get a good night's sleep, eat a wholesome meal and drink water before you take the ASVAB. If you do poorly on the test, schedule to retake it.
6. Do not be "talked into" a military job (MOS) you do not truly want. Also, make sure the MOS you enlist for is written properly in your contract. A simple typographical error could have long-term consequences.
7. Try to schedule your basic training in the early spring or early fall months. This is important because the extreme heat of the summer, or cold of the winter makes training even more miserable.
8. READ YOUR ENLISTMENT CONTRACT BEFORE YOU SIGN IT! MAKE SURE ANY PROMISES OR

BONUSES ARE CLEARLY WRITTEN INTO YOUR CONTRACT.

9. Make clear copies of all documents you will ever be given by the government. Keep your official documents in a safe and accessible area. Do not give the military your actual birth certificate, or other important documents—give them copies.
10. Try to sign up for a desirable first duty station of choice—the first place you will live after basic training. However, if you enlist for more than four years, chances are you will be shipped to areas where the military needs you.
11. Never be pressured to sign your contract, or anything you do not want to sign. If you think something is legitimately wrong with a document, then you are under no obligation to sign it even though you may feel highly pressured to do so.
12. Check into the Reserves/National Guard to see if these components of the military can meet your needs with fewer obligations than active duty service. Enlist for the shortest time possible if it suits your needs.
13. Once you sign your enlistment contract and take the oath (swear in), do not look back. For sound mental health, develop a positive attitude and look forward to your new military life. Start making preparations for basic training—read the next chapter thoroughly.

Chapter 3

Second Step
Golden Rules of *Pre*-Basic Training

These next chapters are specifically devoted to readers who have already enlisted. The most important parts of this book are given in order of usefulness for those recruits actively preparing for basic training. However, if you have not yet enlisted, this chapter might give you a feel of what to expect immediately after you have signed your military contract. If you have just signed your enlistment contract, you usually have between three and nine weeks to prepare for basic training. In any case, you do not have much time and must be exceedingly serious about your pre-basic training preparation.

Immediately After Enlisting

As was outlined in the previous chapter, enlisting in the military is a somewhat lengthy process. Very rarely can anyone sign up for the military and leave for training the next day. You must first meet with a military recruiter, take a two to three hour aptitude test, undergo a medical examination and have your background checked. Records and paperwork must be carefully reviewed. If you pass all the prerequisite criteria and are allowed to sign your official enlistment contract, most likely you will be sent home from the military entrance processing station (MEPS) for a few weeks to prepare for military training.

It is quite common to experience a range of emotions immediately after enlisting. Some recruits feel excited, while others may feel anxious. After your emotional state equalizes, do not relax too long because you must start taking serious action to prepare for basic training. Ideally, you should have a little longer than six weeks to actively condition yourself for boot camp. You absolutely must use your time wisely to ready your body *and* mind for military life. Every day you have before you leave for basic must be effectively utilized either with mental training, or physical conditioning. Keep in mind that it takes a minimum of 30 days to realize any meaningful lasting gain from exercising. Additionally, the more time you have to memorize the required military tasks, the more likely the information will endure in your memory and become second nature.

Here is the drill for this chapter. You need to read the next few pages thoroughly and understand it. This chapter is not meant to delve into great detail, but serves as a general preparation outline. Recruits' pre-training will be different due to their life style and available time. A positive aspect of physical conditioning is that a pre-training regimen does not need to be very time-consuming. If you dedicate just one hour a day to your plan, that can be a solid foundation to start actual basic training in a few weeks. Unfortunately, most people will start a good plan only to drop it after a week or two. My advice is to devise your training schedule so it is not too difficult and stick to it every day! Your time is extraordinarily valuable now that have enlisted, so DO NOT WASTE IT.

Pre-basic Training, What You Need To Do

Assuming at least six weeks before shipping out for basic training, the following items are your focus points.

1. **Get at least eight hours of sleep every night**. Start getting in the routine of going to bed before 10:00 o'clock p.m. (2200 hours in military time). Start waking up at least by 5:00 o'clock a.m. (0500 hours). Your body has an internal alarm clock and must get set to a new routine.

Insider tip: The human body needs at least four weeks before something becomes a habit. If you are not in a routine of going to sleep early and waking before the sun, you will have adjustment problems in boot camp. Also, if you are actively exercising, your body rebuilds itself at night while you sleep. You must give your body time to rebuild, or you will only get weaker. Sleeping and resting are just as important as exercising--do not underestimate them.

2. Start eating healthy and drink large amounts of water. Substantially decrease your intake of candy and sodas. Start increasing your intake of water and eating a balanced meal. You should drink two full glasses of water per day—more is better. This does not sound like much water, but research shows a large percentage of people do not even drink one glass of pure water a day. Water cleanses the body of impurities and helps moderate the negative effects of exercising such as cramping and dehydration. Besides, if 75% of your body is water, what do you think you need the most?

As far as food intake goes, I am not advocating cutting out all sweets, but just start modifying your diet to include things which are really good for you like vegetables and high quality meats. Hot dogs, fried foods, fast foods, chips and most snack foods are not good for you and should be eliminated from your diet. You will not have access to any type of junk food during basic, so get your body used to it. Remember, your body is comprised of the type of food you put into it. The lower the quality of food, the worse you feel.

3. Break your addiction to tobacco. When I went through infantry training, I saw several people hurting for tobacco. Saying tobacco is not addictive is like saying drill sergeants like being called "buddy." If you dip, chew, or smoke tobacco you will be craving it severely while in training. You must earnestly try to stop using tobacco products. If you need help, get it because the drill sergeants will not be sympathetic to your cravings. If you get caught with any tobacco product while in basic, you will pray for lethal injection.

4. Start Running Now! You do not have to be a track star, or even like running to excel in the military. However, there are three major types of timed exercises you must do for your physical test (PT);

a. Push-ups
b. Sit-ups
c. A two-mile run (three miles for Marines)

The human body has the amazing ability to adjust to many difficult physical activities, but you must give it a fair amount of time to get used to it. From personal observations, I saw how difficult it was for people in my platoon who did not run before they arrived at training. Out-of-shape recruits were sent to special "fat boy" training sessions. These special sessions define the term pain---trust me, you do not want to go to these special workouts. The last thing a new recruit needs is having to adjust to extreme soreness of leg muscles repairing themselves. If you can run two or three miles every other day, you will be able to handle the first stages of basic because you will not be as sore or fatigued.

Let's say you have six weeks before shipping out and you are somewhat athletic. I would start jogging a half-mile every other day for the first week. The second week I would jog one mile every other day. The third week, try two miles; the fourth week try three miles. By the fifth week if you are running two to three miles with ease every other day, you can make it in basic. When you do your two-mile run during basic training, you will be timed. Depending on your age bracket, you will need to complete the run within a very specific time frame. For most people, they should run two miles within 15-16 minutes. Technically, males between the ages 17-21 must complete a two-mile run in less than 16 minutes 36 seconds. Males between 22-26 must complete the run in 17 minutes 30 seconds. Females 17-21 complete the run in a maximum time of 19 minutes 32 seconds.

Insider tips: You need to buy running shoes while in basic training. Your shoes must be running shoes, not any other type like cross trainers, or basketball shoes. If you do not have a fairly new pair when you enter basic training, the sergeants will make you buy a new pair of shoes at the post exchange store (PX). The PX does not have the selection of civilian stores. Additionally, you will have to pay for running shoes with your money---the military does not buy athletic shoes for you. My advice is to spend the money to buy a good pair of running shoes before you ship out. Do not skimp, but you can buy a good pair of "mid-mileage" shoes for around $60-$70.00 dollars. If you have narrow feet, Nike is good. Asics and New Balance are good for wide to medium wide feet. You do not have to buy top of the line "high mileage" shoes unless that is what fits you best. What is most important is that your high quality running shoes fit your feet very well. Many $120 shoes do not fit as well as $65 shoes, so try them on in the store and walk around. Make sure you have enough room in the toe box area of the shoe and that it has good shock absorption.

Another good tip is to understand how you run. Some runners start out slow and speed up during their last mile. Other runners have a steady pace throughout their timed distance. It is important that you understand your running style and stick to it during training. I vividly remember my first physical test (PT) during basic. The drill sergeant said "go" and immediately half my platoon shot passed me. Some recruits even chided me for being slow. However, I fully understood my pace and maintained my style of running. After the first mile, I passed all of the recruits who seemed to be severely out of breath. My time was the third best not because I was a great runner, but because I knew *how* to run. Bottom line is to pace yourself and do not give into sprinting.

5. Start doing push-ups. Another exercise you must do for your physical record and that is also the darling exercise of the drill sergeant is the venerable push-up. At a minimum, you should do at least 30 push-ups in two minutes. Even though this is a strictly timed test, form counts just as much as the amount you must do. Your back must be straight, feet must be together, head up and you must lower yourself to the ground so that you are parallel to the ground without arching your back. Most people know how to do one good push-up. However, you must do each push-up just like the first one.

Insider tip: Do not bounce your chest off the ground when you are doing your push-ups. Also, do not spread your hands out too far from the sides of your body. Drill sergeants love to catch those people who spread their arms out so wide they only have two inches between their chest and ground. Your arms should be just a little more than shoulder-width apart. Keep your head up slightly, but take precaution against injury. Many neck and back strains come from this exercise. The best way to prepare for this event is to slowly build up the amount you can do with the proper form. Starting from not doing any push-ups to 25 per day only takes about three weeks. By week six, you should be able to do around 35 push-ups without a struggle (for males). Once you reach a solid degree of arm and chest strength, devise a set approach to do them. For example, I would knock out twenty push-ups at a time, then rest for four slow seconds. You are allowed to rest, but you must do so in the "up" position with both hands and feet on the ground. Knowing when to rest during when doing push-ups is critical.

To pass a physical test, males between 17-26 should be able to complete about 35 push-ups within two minutes. Females in the same age bracket should complete about 13 push-ups.

6. Learn to do sit-ups correctly. I hated these back killers more than any exercise. The military would be well advised to do away with sit-ups due to the injuries incurred by them. I do not know of any physical trainer who endorses this old school exercise, yet it persists in the military as of this writing. Unfortunately, until the act of congress is passed to change the sit-up requirement, you should to do at least 30 of them in two minutes. (Like the other exercises, consult a recruiter for your age bracket requirements). The focus here is to be mindful of your back and protect it. Most recruits will start out with their knees bent, someone holding their feet down and their hands behind their head. The injuries happen when recruits try to "crunch up" too fast and fall back hard on the small of their back. Sometimes, recruits hurt themselves when they pull their neck and head up with their hands.

My advice is to train on a foam mat, which you are allowed to use in the military. Do your sit-ups at a moderately quick pace, but be mindful of your form and back. Do not strain anything but your stomach muscles when crunching. Do not come down hard on your back and watch your head. Remember to breathe! Most recruits hold their breath on the up-take and do two or three sit-ups without breathing regularly. You need to get maximum oxygen to your muscles at all times during exercises. Last, you are allowed to rest in the "up" position—resting with your back straight and your head near your bent knees. As with the push-ups, strategically resting after 20 or so sit-ups might help your endurance. Do 20 sit-ups, rest four seconds, and do 20 more. Repeat this routine until you cannot do anymore.

For the record, 17-26 year old males should complete about 47 sit-ups in two minutes. Females should also complete about 47 sit-ups in two minutes.

7. The Big Three Routine: As mentioned, the big three PT exercises are push-ups, sit-ups and running. You should do the exercises in the prior order. Previously, the military made you run your two miles, and then you did your timed push-ups and sit-ups. What the military trainers discovered was that PT scores improved if they allowed personnel to do their timed run last. The energy drained from the body after a two-mile run was substantial enough to decrease relevant muscle groups needed for other exercises. My advice is to train with push-ups first, sit-ups next, and running being the last type of exercise.

8. Extended Training and Stretching: If you want to effectively meet the minimum standards during basic training you should train beyond your minimum score level. Certainly, do not rush to build up muscle, or overtax yourself. However, training beyond the minimum standards affords you a margin of comfort, at least for the first two phases of basic. For example, if you need to do 35 push-ups in two minutes for your age bracket score, you really should be able to do about 45. In my opinion, you should do 15-25% more than what is required of your exercise age bracket.

Another suggestion about conditioning involves stretching. Stretching your muscles is an important part of any exercise plan. Although, a few physical trainers take issue with stretching, you will have to do stretching exercises during morning PT. I am not a physical trainer and cannot give advice on how to stretch or train. However, stretching has always helped me. I suggest you speak to someone who is qualified to show you a few practical stretching exercises.

Insider tip: Keep in mind that one of your primary goals during basic training is to fight fatigue and soreness. Initially, your body will see exercising as something alarming and wrong. Starting an exercise program will severely alter your sleeping and eating patterns. Lactic acid is produced when exercising and muscles will be broken down. It is imperative you give your body time to adjust to rigorous activities. Again, you will NOT have the luxury of adjusting to fatigue and soreness during boot camp. You must give your body adjustment time *before* basic, not during. Simply put, the more exercise you do beforehand, the easier time you will have at boot camp.

It is well known that proper stretching, drinking large glasses of water everyday and sleeping eight hours per night makes physical adjustment substantially easier.

9. Start reading the IET book. Outside of vigorously starting an exercise program, studying your IET book is extremely important. Think of it as exercising your mind. It may seem too simple, but I have seen where recruits had the opportunity to get ahead by reading their IET book and did not. Having an IET book is like getting the answers before a final exam. Study it, learn it, memorize it! Reading a few pages in the comfort of your bed before going to sleep is far more effective than trying to memorize a certain page when a 235-pound crazed drill sergeant is screaming obscenities at you.

If you have several weeks before shipping out, read the IET book and familiarize yourself with its content. The information is not hard to learn and is quite elementary. About two weeks before shipping out, I would start memorizing the parts which are outlined in later chapters.

Insider tip: From your IET book, it is usually best to memorize the military alphabet, the general orders, how to report troop movements (SALUTE), how to use the radio, core values and the chain of command.

10. Work on your attitude. Perhaps more than any other suggestion outlined in this book, developing a positive attitude is the most important factor for success. Although this point seems like common sense, countless recruits fail to work on their attitude before they arrive at boot camp.

A good attitude not only leads to increased life satisfaction, but can literally save your life. For example, research has illustrated that the difference between prisoners of war who survive versus those who died is that survivors owed their life to having an optimistic outlook on life.

It appears most people's attitude is predicated on their immediate environment and circumstances. The stressful atmosphere of basic training is not a place where you want to be taking cues on how to develop your attitude. A good attitude must come from within a person and must weather the events which take place externally.

How does one develop a good attitude? My belief is that good attitudes stem from having a clear sense of purpose and ardently focusing on the positive aspects of events. Knowing who you truly are, including your strengths and weaknesses, is important to understand before training. For example, I remember a person in my squad told the drill sergeant that he was a pacifist. Obviously, this recruit did not take a thorough introspective look at himself before he enlisted. Needless to say, training was very difficult for him because the military is not nurturing of pacifists and the drill instructors tried to weed him out.

Insider tip: Part of developing a good attitude about basic training is to clearly understand what you want out of it. My advice is to literally write down three good things you want to derive out of training and the military. Bring these goals with you to training and look at them when it gets tough for you. For example, my goals (besides serving my country with honor) were to get into the best shape of my life, get money for college and learn leadership skills. I put my goals against the backdrop of the adage, "Whatever doesn't kill me only makes me stronger." Trust me, this goal theory works. Know what you want, go after it, and do not look back with regrets.

Other helpful training tips:

Although not necessary, it is helpful if you lift weights before basic training. I do not recommend getting bulked up, but work on building endurance. People mistakenly believe that boot camp turns them into body builders; it doesn't. The military emphasis is on strength and endurance, not body sculpting. Extra muscle mass is tough to carry around during long training exercises. While big muscles are important, the strength associated with endurance is better suited for training.

You might also prepare your feet and legs for basic. Wearing boots and toughening the soles of your feet might be beneficial. While in basic, it is said recruits march about 100 miles (in total) wearing boots. Unfortunately, the boots the military issues are not like the high-tech hiking boots you buy at an outfitter. Your military boots have all the comfort of plywood.

Another task which is helpful to learn is marching. Although I was a solid runner, I hated marching and drill. Learn how to march correctly and the basic drill commands such as, "right face." Your IET book (also see the Appendix) has specific details.

Practice "flutter kicks" and leg lifts. These two exercises are common exercises used to motivate (punish) trainees. Here is how to perform these exercises: Lay on your back with your hands under your butt. Lift your feet and legs about six inches off the ground while keeping your legs straight. You must hold this position for at least 30 seconds, sometimes a minute. The flutter kick is done the same way as the leg lift, but with the addition of moving your feet in an up-and-down motion (like scissors).

Last, once in a while before you ship out, visit a park to climb on the monkey bars, or gym set. Also, throw baseball-sized objects as far as you can. These exercises are helpful when you encounter the obstacle course and throwing grenades during training.

A Pre-Basic Training Schedule

The following training regimen is only an example of how to prepare for basic training. Some people might have more time to devote to training, while some people might be shipping out sooner than others. Also, this example is for people who have not exercised in a long time, or are just starting an exercise program. I know this is a very simple and easy example, but it worked effectively for me. My advice is to have at least six weeks before leaving for basic. The longer you have to pre-train the better. You should tailor your pre-basic schedule to fit your lifestyle, personality, and interests.

Before you start your plan, heed this sound piece of advice—be mindful of burn out. **DO NOT** rush into a full-blown exercise plan. Start slowly and gradually to build up your mind and body. The closer you get to your ship out date, the closer your lifestyle should mimic a true basic training schedule. However, you MUST give yourself time to adjust.

Example Pre-basic training program:

This schedule may seem simplistic, but it works and safeguards against burn out. Make sure you rest properly where it is noted. Also, women should do half of the stated push-up exercises.

Week I

Monday

Morning: 10 push-ups, 10 sit-ups, jog a 1/3-1/2 mile. Time 25 min.

Night: 10 push-ups, 10 sit-ups, walk a ¼ mile. Time 25min.

Read ten pages from the IET book (easy reading).

Tuesday

Morning: 10 push-ups, 10 sit-ups, jog a 1/3-mile. 25 minutes

Night: 10 push-ups, 10 sit ups, walk ¼ mile. Time 25 min.

Read ten pages from the IET book.

Wednesday

Morning: 10 push-ups, 10 sit-ups, jog a 1/3-mile. 25 min.

Night: 10 push-ups, 10 sit ups, walk ¼ mile. 25 min.

Read ten pages from the IET book.

Thursday

Morning: 10 push-ups, 10 sit-ups, jog a 1/3-mile. 25 minutes

Night: 10 push-ups, 10 sit ups, walk ¼ mile. 25 min.

Read ten pages from the IET book.

Friday

Morning: 10 push-ups, 10 sit-ups, jog a 1/3-mile. 25 minutes

Night: 10 push-ups, 10 sit ups, walk ¼ mile. Time 25 min.

Read ten pages from the IET book.

Saturday-- Take the day off, but read the IET book at night.

Sunday

Morning: 10 push-ups, 10 sit-ups, jog a 1/3-mile. 25 minutes.

Night: 10 push-ups, 10 sit-ups, walk a ¼ mile. Time 25 min.

Finish reading the IET book.

Week II

Monday

Morning: 15 push-ups, 15 sit-ups, jog ½ mile

Night: 15 push ups, 15 sit-ups and read IET book.

Tuesday - Friday-- Repeat Monday's schedule.

Saturday-- Take the day off. Review your IET book.

Sunday-- Same as Monday

Week III

Monday

Morning: 20 push-ups, 20 sit-ups, jog 1 mile.

Night: 15 push-ups, 15 sit-ups (no reading)

Tuesday

Same as Monday, but do not jog.

Night: Start re-reading IET book.

Wednesday--Same as Monday

Thursday--Same as Tuesday

Friday--Same as Monday

Saturday --Day off

Sunday --Same as Monday

Week IV

Monday

Morning: 25 push-ups, 25 sit-ups, jog 1.5 miles.

Night: 20 push-ups, 20 sit-ups

Tuesday

Same as Monday but do not jog. Build a positive attitude.

Wednesday--Same as Monday, but skip morning push/sit-ups.

Thursday--Same as Tuesday. Think about 3 personal goals.

Friday--Same as Monday, but skip morning sit/push-ups.

Saturday--Day off. Write your 3-military/personal goals down on paper.

Sunday--Same as Monday

Week V

Monday

Morning: 30 push-ups, 30 sit-ups, jog 2 miles.

Night: 20 push-ups, 20 sit-ups.

Tuesday

Morning: 25 push-ups, 25 sit-ups, do not jog.

Night: Start memorizing the necessary parts of the IET book.

Wednesday--Same as Monday

Thursday--Same as Tuesday

Friday--Same as Monday, but only do the night push/sit-ups.

Saturday--Day off

Sunday--Same as Monday

Week VI

Monday

Morning: 40 push-ups, 40 sit-ups, jog 2.5 miles.

Night: 20 push-ups, 20 sit-ups.

Tuesday

Morning: 30 push-ups, 30 sit-ups. Do not jog.

Night: Memorize and reread parts of the IET book.

Wednesday--Same as Monday

Thursday--Same as Tuesday

Friday--Same as Monday

Saturday--Day off

Sunday--Same as Monday

***Note**: Each week includes getting the proper eight hours of sleep, changing your diet to include wholesome foods, and substantially increasing your water intake. Also, you should be decreasing your junk food intake. Be mindful of Week IV. Take care in creating a positive attitude very seriously. Know what you want to get out of basic training and remain focused on your goals for the next three months.

The Farewell Party—Don't Do It

Some family or friends throw a big party for their departing loved one. While having a party is a nice gesture, I strongly advise recruits not to overdo their celebrating. Getting drunk, smoking pot and getting high is especially stupid the night before shipping out for basic training. There is certainly nothing wrong with having a beer, eating party food and dancing; just do things in moderation. I cannot emphasize enough how important your rest and sleep are the night before you leave. You should get at least seven to eight hours of good sleep the night before training—you will need it.

If you have a party thrown in your honor, plan it two days before you ship out. This way, you will have a day to recuperate and still get a good night's sleep before you depart.

Chapter Summary

1. Start training your body and mind as early as possible before shipping out
2. for basic training. Six to eight weeks is usually enough time to prepare.
3. Secure an Initial Entry Training handbook from your recruiter and study
4. it very carefully. Review it several times every week.
5. Get at least eight hours of sleep every night. Fall into a routine of waking up early in the morning. Remember, rest is just as important as exercising
6. Start eating healthy and drink large amounts of water. Drink at least two to three full glasses per day (six to eight glasses is ideal).

7. Break your addiction to tobacco if you have one.
8. Buy a good pair ($50--65.00) of running shoes and start running.
9. Start doing push-ups. Remember to work on the correct push-up form.
10. Start doing sit-ups. Work on the correct sit-up form.
11. Exercise in this order: Push-ups, sit-ups, and run. Order is important.
12. Stretching is important before exercising. Find out how to properly stretch your muscles before you begin any exercise. Never bounce when stretching and stretch slowly.
13. Immediately before your ship-out date, you should be able to go beyond your minimum physical exercise requirements. Focus on your weak areas of physical conditioning.
14. If your family or friends have a departing party in your honor, plan it a few days before leaving. It is acceptable to celebrate, but not to excess. Remember, all recruits will be screened for drugs.
15. Work on your goals and attitude. You must develop a positive attitude.

Chapter 4

The Basic Training "Commandments"

Without question, this chapter is the most important concise chapter in this book. I strongly encourage all readers to study this chapter carefully so they may have greater insight about military training. This section is purposely short because you will want to memorize many of the points. New recruits need to review this part many times immediately before they arrive at their training base. Knowing what to expect alleviates much of the anxiety associated with basic training.

I call the following list "commandments" because they are the most important items to remember which may help you stay out of trouble during training. These commandments might make your induction into the military a little less stressful. Note how I said a little less stressful, not stress-free. No book by any authority can tell you how to make boot camp stress-free. This guide, and in particular this chapter, only helps to make the odds better that basic will be easier, not easy.

As the introduction mentioned, you must be aware of the situation in which to utilize these commandments. In other words, use common sense. For example, although I advise recruits not to volunteer for tasks, if the drill sergeant is looking at you while asking if anyone wants to carry the M-60 machine gun, you do not look away. What I will adamantly state is this; do not break any rules, adhere to all safety procedures, be a good team player, and have a positive attitude.

Your primary goal during the first few weeks of basic training is to adjust to a strict military way of life. After all, you are being indoctrinated and torn away from a softer civilian life. By following these commandments, attaining your goals should be within your grasp. Do not worry if you do not fully understand the rationale behind many of the commandments; most will be detailed in the following chapters.

The Basic Training Commandments

1. **Do not draw attention to yourself while in basic training**. From the time you arrive at your training base until the time you leave, keep a low profile. For example, wear normal clothes when you arrive and have a neat clean-cut appearance. Do not wear conspicuous tee-shirts, have long hair, or an unshaven face. Try to blend into the crowd as soon as possible. The longer the drill sergeants do not know your name the better for you.
2. **Try to position yourself toward the middle of your formation.** In my opinion, you want to be standing about three or four positions in from the end of your squad. Generally, you do not want to be the person who is in the front row, or at the end of each line. Sometimes the drill instructors know the middle of the platoon is where recruits like to hide from them. Therefore you may not want to be exactly in the middle, but just near the center out of the sergeants' direct line of sight.
3. **Study your IET book every chance you get**. As incredible as this sounds, there will be times during basic training where you will have "down time." Down time is not a synonym for "free time," but rather a few extra minutes with nothing to do while waiting for a task, or an event. There are parts of your IET book (which will be outlined in this guide) that you must <u>memorize</u>. My advice is to over-memorize whatever needs to be learned during down time. You will be thankful you over-memorized the IET book when the drill sergeant is yelling so loud you forget your own name. Some things which

need to be memorized are the military alphabet, general orders, core values, rank structure and the names of your commanders.

4. **Do not volunteer for anything—at least initially**. I am not advocating that recruits shirk their responsibilities, or hide from their assigned duties. Volunteering for military duties and tasks is extremely important *after* you complete basic training. However, *during* basic training, you will receive no extra money or time off for any extra work you do. Volunteering for tasks only puts you at risk for more physical exertion, missing valuable sleep and for the drill sergeants to know your name. Once you volunteer, the more likely the drill instructor will call upon you again. I have seen recruits volunteer with good intentions only to be punished for not doing the job they volunteered for well enough for the drill sergeants. Remember, you will never please a drill sergeant until the final three weeks of training.
5. **Always keep in mind that you will never please a drill sergeant**. Perhaps "never" is an exaggeration, but for the first 75% of your training cycle you cannot do any military task up to your drill instructor's satisfaction. This is the "game" the sergeants play to let you know who is in charge. You must always try to complete all assigned tasks as best you can, but realize you will probably get yelled at for not doing it right—even though you completed your task perfectly. Just go with the mind game and do not get frustrated.
6. **Do not try to "buddy up" to a drill sergeant**. I cannot overstate this commandment. I have seen where some drill sergeants purposely try to be nice to a recruit, only to set them up and slam them. My advice is never make any type of casual comment to a drill instructor. Do not speak

unless spoken to first. Make your responses to drill sergeants very brief. Try not to isolate yourself with a drill instructor. Never, never, never, touch a drill sergeant, or call him just "sergeant." Mostly, address your drill instructor as "*drill* sergeant ______." This commandment might be easy to follow during the beginning stages of training, but during the last few weeks you will be lulled into a false sense of security. Be very cautious of "nice" drill sergeants; chances are they are playing mind games.

7. **Build your mental wall and do not take anything personally**. Almost everyone during basic training will have a drill sergeant yelling obscenities in his or her face. When you find a 250-pound red-faced sergeant screaming so close to you can see his facial veins popping, you must let his/her words bounce off of you. Even if the drill instructor is calling you and your mother a "shit-eating mofo" you cannot take it personally. This is all part of the basic training game and is designed for the drill sergeants to control you and to see if you can tolerate stress. Visualize the drill sergeants' words bouncing off of you as if you were a brick wall and the words are rubber balls. Remember, basic training is much bigger than you or any individual.
8. **Drill sergeants lie to you**. Drill instructors do not lie because they are liars, but lie because it is part of the basic training game. Drill sergeants want to assert their control and cause frustration in new recruits. If a sergeant says you can use the telephone after completing a task, then says you cannot after you have done the task, blow it off. However, if you become upset and show your frustration by the slightest smirk, then the drill sergeants have accomplished their goal and will rip you up.

9. **While standing in formation at the "position of attention," DO NOT MOVE.** This is the drill sergeants' favorite time to catch recruits doing something wrong. Drill sergeants love to administer punishment exercises. It is almost as if the sergeants are in competition with one another to see how many recruits they can catch and punish. One of the first things you will learn in basic training is how to stand in formation. You will learn how to stand at the "position of attention." Once you learn how to stand, be extremely still. This means do not move your eyes, your nostrils, your hair, your fingers, toes---anything! Even if a wasp is taking a chunk of skin off your neck, do not move while at the position of attention.
10. **Never show emotions**. There will be a range of emotions you will experience during basic training. Your emotions will range from fear, sadness, hopefulness, and even happiness. However, you must never show the drill instructors your emotions through your facial expressions, words, or non-verbal body language. Showing emotions when you are stressed out or frustrated is the worst type of emotion to show a drill sergeant. You will pay dearly in the form of doing push-ups until muscle failure if you show your frustrations. Showing signs of anger or stress will be taken as a sign of weakness by the sergeants. However, even when you feel good about something, hide it. This is not to say you cannot be yourself around your platoon members. However, when the drill sergeants are around, you better be an emotional ice cube. The keystone term to remember with this commandment is "military bearing." Sergeants love to say, "Never lose your military bearing."
11. **Lock up your gear and watch your possessions**. I have been criticized for listing some points which are obviously common sense. However, at least 15% of the recruits in

my platoon had items stolen from them. I even had a good friend lose a bankcard-- and the thief stole money from his account. I am not advocating paranoia, and you must trust people in your platoon, but it is always wise to lock up your belongings (and I do mean lock with a good lock). I can confidently say that people in your platoon will get something stolen from them. It happens all of the time. Besides your money, make sure you keep your military identification secure. Your ID is critically important and you will be in a world of trouble if you lose it.

12. **Watch your weapon and memorize all numbers on it**. During weapons training, you will be issued a weapon, most likely a M-16A2 assault rifle. If you "misplace," or lose this weapon you will seriously wish you never enlisted. Besides hitting a drill instructor, I cannot think of another offense which can get you in more trouble than misplacing your weapon. When you check out your weapon from the armory, always keep it in your hands, or on your body. Unless directly ordered to leave your weapon, always make sure you can see and touch it. There will be two numbers on your weapon you must memorize. The first is a serial number which is engraved on the weapon by the manufacturer. The second number is the weapon number assigned by the armory sergeant at your base. The weapon number is usually shorter than the serial number. If you cannot repeat both numbers back to the sergeant as fast as your own name, then you will be doing punishment exercises until muscle failure.
13. **Be extraordinarily safe and vigilant at all firing ranges**. Like losing your assigned weapon, if you screw up at the firing range you will hate life. The military takes safety at the firing range extremely serious. There are very strict ways of doing things at the range, and you must do things

in tightly structured steps. Listen to the range controller's instructions. He will walk you through each step, but you must always pay close attention. After several trips to the firing range, some recruits get too relaxed and comfortable. Usually when people become too comfortable doing a task, mistakes happen. You never want a mistake to happen on a firing range. If one person causes an error during a live fire drill, the whole platoon could get seriously punished because of that person. Unless you enjoy being the target of your platoon's hostility, I would not make a mistake at the range.

14. **Always leave your weapon's safety switch in the "on" position.** I know this is another common-sense item. However, when you are dead tired from the relentless exercising and lack of sleep, it is easy to forget to make sure your M-16's safety is in the "on" position. I saw a drill instructor actually grab a recruit around the neck because his safety was in the "fire" position while the recruit was leaving the range. Even if your weapon is not loaded, or doesn't have an ammunition magazine inserted, you must be mindful of the safety switch.
15. **Always point your weapon downrange and away from any person.** Even if your weapon is not loaded, never point it anywhere but at an assigned target, or at the sky. If your weapon even appears to be pointed in the general direction of others, you will get "smoked" beyond imagination. Remember, the killing range of an M-16 is almost two miles!
16. **Make sure your weapon is empty and you have no ammunition when leaving the firing range.** Do not double check, but triple check that your weapon is empty and free of ammunition when leaving the range. Sometimes bullets become stuck in a chamber, so check the

chamber visually. Sometimes extra bullets end up in your uniform by accident. Sometimes recruits try to take an M-16 bullet home. Do not take even an empty shell back to the barracks. You not only risk being kicked out of the military, but serving jail time if you steal ammunition. When leaving the firing range, you will have to yell out, "No brass, no ammo drill sergeant!"

17. **Take advantage of hidden down time**. Many times you will experience the "hurry up and wait" mentality of military training. The drill instructors usually cannot keep every recruit busy, or monitor you at all times. My advice is when you have completed a task, or do not have a task to do, blend into your squad and rest. Sometimes when I just came off the live firing ranges, and was cleaning my weapon, there was ample time to rest due to the large volume of soldiers still needing to shoot. While resting, you still must be prepared to look busy.
18. **Economize your motion**. During most of basic training, you will be severely sleep deprived, fatigued and just plain worn out. It is important to find times to rest without being caught (as mentioned in #17). Do not volunteer, or do any unnecessary activities which will drain your limited supply of energy. Conserve your strength when it is appropriate.
19. **Look busy even when you are not.** While it is important to rest during unofficial down times, do not let the sergeants catch you sleeping, or relaxing. You can appear to look busier than you actually are—especially when you know the sergeants are watching you. For example, in my training company, we had to always *run* everywhere we went and act like we had a purpose doing something at all times.

20. **Do not be a loner**. No matter how tough or bold soldiers are, they still need to have good social support. I won't bore you with the psychology behind having close friendships, but humans need them to stave off stress. Try to make at least two good friends you can talk with and count on during training. Don't try to make friends with everyone, but having a few solid friends is extremely important.
21. **Use last names only**. Never call friends, or anyone by their first name during training. Use only the last names of people. Training is supposed to be very formal. While the DI's know you are making friends, they want you to be strictly business in front of them.
22. **Teamwork, teamwork, teamwork**! From day one, your platoon must unofficially decide who can take charge. If you have squad leaders assigned early, make sure all duties (sweeping, mopping, restroom detail) are assigned to specific people. The *earlier* you have a team climate, the easier your drill instructors might be on your platoon. Do not ever let your drill sergeants hear any type of fighting, yelling, or bickering among the recruits—be a team!
23. **Watch what your platoon says and does.** Keep negative comments to yourself and tell your buddies to also watch what they say. Keep horseplay to a strict minimum. Any negative statement which is overheard can be used by the sergeants to punish the platoon. For example, after a long road march my platoon went back to the barracks and started complaining about the heat. Unknown to us, the sergeants listen to everything we "bitched" about over the intercom. Needless to say, our platoon did not sleep that night due to a marathon of punishment exercises.
24. **Do not fall asleep during guard duty**. Sometimes at 3:00 in the morning when two recruits are on guard duty,

the drill sergeants may hold a surprise inspection to see if the guards are asleep. If you want to see true unabridged rage from a drill instructor, let him/her find you asleep when you are assigned guard duty. You will find staying awake in general is very difficult during training. You will be sleep deprived most of the time; therefore you must be very strong during guard duty. Do whatever it takes to remain alert, but do not give in to sleeping when your platoon is counting on you.

25. **Do not fall asleep during boring military activities and tasks**. From the time you arrive until you leave, you will probably be deprived of sleep. Sleep deprivation is a form of control and the military utilizes it frequently. Most new recruits will be most fatigued during the first two weeks of training. During this time frame, you will struggle to keep your eyes open throughout the day. One favorite trick of the training commanders is to haul a company of recruits together in a relatively small meeting room. The room will be overcrowded and very warm. This is possibly done to induce drowsiness. While watching very boring videos regarding some military activity like dental hygiene, recruits will fall asleep. The drill sergeants will be circling like hawks watching for prey. Do not fall asleep anytime, or anywhere you can be caught.
26. **Positive Mail Call**. Tell your friends and family they must write you letters while in basic training. Moreover, all letters must be positive and cheery. I cannot describe how or why mail call is so very important, but you need the connection letters bring. I have witnessed incredibly strong soldiers become visibly upset because they were the only ones who did not receive a letter from home. Also, I was privy to an attempted suicide by a recruit whose insensitive girlfriend broke up with him during ba-

sic via a letter. Again, be positive in your writings back home and have friends send good letters back. Also, do not have friends send you pornographic pictures, contraband items, or food through the mail. Such items are quickly screened and taken.

27. **Buy and bring extra black boot polish**. You will spend a long time cleaning and polishing your boots. Buy high quality black paste wax *and* liquid wax. Paste wax last longer, but the liquid quick wax is valuable when you forget to polish your boots. Kiwi brand is good and makes both kinds of wax. I would bring at least three extra containers of high quality wax.

28. **Buy extra Q-tips and pipe cleaners to clean your M-16**. Cleaning your weapon is a frequent task and you will run out of weapon cleaners quickly. Buy the highest quality Q-tips and pipe cleaners available.

29. **One to show and one to wear**. After breaking in both sets of your boots, keep one pair highly polished for inspections. A few times, switch wearing your boots, but always have one in good form. Also, your everyday uniform (BDU) should be worn in a similar fashion—one set very clean and neat for inspection, and one to wear. You can usually get away with wearing your BDU's at least twice, if not three times before you need to wash them.

30. **Always keep clean**. Not only should you keep clean and neat, but your laundry must be done on a regular basis. If you produce odors (stink) because of your clothes, or poor hygiene, the DI will smoke you. This commandment is easy to follow during Phase I because of the DI's supervision. However, during later phases of training you will be responsible for your laundry and cleanliness. Try to do laundry promptly during "down time." Do not wait for others to do it first. Be careful not to let laundering your

clothes interfere with your sleep time. Manage your personal time wisely.

31. **Always maintain a neat personal area**. If you consistently keep your bunk bed and locker orderly, you will find it is easier on you in the long run. Sergeants love to hold surprise inspections of recruits' billets when the platoon is outside. Many times, recruits come back to a torn up barracks because they did not keep it up to specifications from the start. Sometimes, sergeants walk around individual bunks when recruits are outside to assess the recruit's level of maturity.
32. **Stay "mid-pack" initially and do not show off.** As with the "blend in" commandment, this point suggests that you do not draw attention to yourself during activities. For example, if you are an excellent runner, resist the urge to show off and stay midpack. In my opinion, physical training is not something you want to show off until the later phases of training. I am not advocating being lazy, or not trying to condition yourself, but you run the risk of being singled out for extra duties if you stand out in the crowd too early. If you want to be a road guard, or something else, that is fine, but you do not receive any extra pay or benefit from extra responsibility during basic.
33. **Do not let your guard down**. Toward the end of training, many recruits start getting careless, or bold. Everyone is excited because training is nearly completed and people are feeling highly confident. However, during these times the drill instructors will set the platoon up to see which recruits will lose their "military bearing." Again, up until you leave the training post, do not joke or relax with the drill sergeants. Do not complete tasks poorly. Some of the harshest punishment exercises came toward the end of

my training. Remember, training is not complete until you graduate, so be careful until that time.

34. **Pack light and bring only essentials.** Many unenlightened recruits pack for basic training as if they were going on a vacation. Do not make the mistake of over-packing. Only bring the minimum amount of personal items necessary. For example, everything I packed for basic training could fit in a small gym bag no bigger than grocery bag. Bring only $40-60.00; light toiletries; extra boot polish; cleaners; and foot care products. Do not bring jewelry, or extra clothes. Remember that you will be issued a huge 50-pound rucksack of military issued (TA-50) gear. Also, you will be able to buy some essentials at the military store (PX).
35. **Keep copies of all your records and military documents**. Always try to keep the originals of your birth certificate and other very important papers in a secure place. The military has a habit of losing records, so make as many copies as possible. If the military gives you any type of document, keep it safe and accessible for years after you leave the service. You will be inexorably sorry if you cannot produce a military document to defend your position on an issue. I cannot stress this point enough.
36. **Go to church**. Even if you are not religious, church allows recruits to escape the relentless yelling which takes place during the first few weeks. Church is one of the few places you can truly relax. You have a right to attend services at least once per week. However, once services are over, make sure you revert back to your recruit mode.
37. **Do not believe rumors**. During every phase of basic training, I heard rumors that our company was finishing early due to overcrowding back at reception battalion, or some other type of wild rumor. Do not believe any rumor

you hear. Make plans to finish basic training in the usual 10 or so weeks. I do not know how these rumors start, but you will hear several during training. Sometimes a drill sergeant may even start a rumor.

38. **Expect the worse, hope for the best**. This is a little psychological trick people can play with themselves to make things seem easier. Many times, if you have an attitude that something will be very difficult (but still have a good attitude to face the difficulty), the activity will actually feel easier when completed. On the other hand, if you believe something will not be hard at all, and the activity actually is very hard, you will perceive it as being extreme. In any event, do not set up your hopes for failure. Be conservative and assume basic training tasks will be difficult. Do not assume much free time, privileges, or any type of rewards—then you won't be let down.
39. **Understand the stages of training**. There are three distinct "phases" of training. <u>Phase 1</u> is the toughest and last about three weeks. During Phase 1, the drill sergeants' job is to intimidate you and indoctrinate you into the reality of military life. The sergeants want to make sure you will follow orders without hesitation. <u>Phase 2</u> is the team and skill-building phase. Phase 2 is a mix of drill sergeant control and military education. Phase 2 is not as difficult as the first phase, but punishments are still common and harsh. The last phase you are given more personal time and chances to demonstrate learned skills. Sometimes during Phase 3 you are treated with respect and you can feel training is nearly complete.
40. **Take care of your feet**. More than any other part of your body, your feet will take the brunt of the tough physical regimen during basic. Take care of your feet, and they will take care of you. Trim your toenails before and dur-

ing basic. Always use a quality foot powder in your boots and socks to keep your feet dry and bacteria-free. Last, buy a high quality pair of boot/shoe insoles. The insoles in the cheap boots you are issued have all the support of newspaper. It is important that you do not buy insoles that are too thick—thin insoles can work fine if they are constructed well. Thick insoles can press your toes against your boots and cause blisters. Check out the wide selection of Dr. Scholl's at Wal-Mart, or other retailers. Try your new insoles on *before* boot camp and see if they fit your feet well.

41. **Pack extra foot care products**. Besides a high-quality foot powder like Desenex, or Gold Bond, you should invest in blister care remedies. "Moleskin" and "Liquid Skin" are excellent products. The moment you feel a hot spot (a soon to be blister), administer the blister care products to the location on your feet.
42. **Keep focused on your goals**. When you arrive at boot camp, keep focused on the last day of training. Keep telling yourself that training is not that long. No matter how difficult training is, it will be over in a few weeks. Remember, you will have some free time after Phase 1 and the yelling will mostly stop during Phase 2. You must give yourself time to adjust—at least a few weeks. Look forward to the fun aspects of training such as shooting your M-16, throwing grenades, or whatever else you may think is cool. Try to be positive, and do not waste energy fighting your situation. Remember your three written goals and re-read them often.
43. **Self-talk**. I promise to keep the psychobabble advice to a minimum, but recruits should realize the power of inner-dialogue, or self-talk. The way people speak to themselves and the words they use can have profound effects

on their attitude and success in life. Quietly, tell yourself you can make it through basic training because others have made it. Do not use any negative words. Always describe any situation as a "learning experience" and describe to yourself the many positive details of the task at hand. On those long road marches, try to carry on long and detailed inner conversations with a girlfriend/boyfriend, or other person. See *humor* in any situation, or whatever else can help you manage stress. Remember, it is your attitude that ultimately determines your altitude in life.

44. **If you are really hurt or sick go to see the military doctor**. You have a right to go on "sick call" and to receive medical attention for legitimate ailments. Even though some drill sergeants will harass you, do not risk permanent injury by being shy, or scared. On the other hand, sergeants know that some recruits attempt to remove themselves from training by faking an injury—do not be one of those recruits because they are usually found out later.
45. **Understand all your rights and obligations *before* basic training**. Many recruits do not understand that when they enlist in the military, they give up a substantial part of their civilian rights. Recruits fall under a different set of laws and rules not found in the civilian world. Study the Uniform Code of Military Justice (UCMJ) for more information regarding recruits' rights. You will be surprised at the differences between military law and civilian law.
46. **Be tough, but civil**. The military is for tough individuals and training is supposed to indoctrinate you into this strict lifestyle. However, a few recruits do have trouble adjusting to the stresses of training. While many recruits manage their stress and are able to complete assigned tasks, some individuals become unable to do things because they

cannot cope with the emotional strain. Unfortunately, these recruits are sometimes singled out for group punishment. Sometimes "hotheads" in the platoon will attempt to gang up on troubled recruits at night. Instead of resorting to violence, it is much more effective to help recruits, rather than hurting them. Do whatever it takes without violence to help each other perform better. Pull together, not apart. Again, teamwork is paramount and it is what the sergeants look for in mature platoons.

47. **Always complete assignments thoroughly.** Your IET handbook should give you an idea of the types of common tasks required of you. If a sergeant assigns a duty to you, or your squad, make sure it is done completely. Always obey orders and rules governing safety.

48. **Do not believe all threats made by your drill sergeants, but be careful.** Many sergeants' threats are empty and used to control recruits through intimidation and fear. However, orders and commands given at firing ranges and those regarding safety must be heeded at all times. Physical contact between a drill instructor and recruit appears to be authorized when safety is an issue. This is why #44 is important to understand. Recruits should know what the drill instructors can, and cannot do. For example, drill instructors cannot physically hurt you with their own hand unless it is for the reason given previously.

49. **Drink water during training.** During hot months, your drill sergeants will order you to drink a full canteen of water frequently. Sergeants will sometimes make sure you obey the orders to hydrate yourself. Recruits have died during summer training due to not properly drinking water when ordered. Although fatalities are very rare, heat casualties can be common in the form of heat exhaustion and heat stroke. Even mild heat exhaustion due to inade-

quate water intake is very painful. One thing drill sergeants highly encourage is drinking plenty of water.

50. **Wear it correctly**. You must know when and how to properly wear your uniform. For example, always remember you must wear your cap/headgear outside. Drill sergeants hide near the dining hall and catch hurried recruits leaving the hall without their headgear. Besides not wearing one's cap properly, another popular reason for drill instructors to "smoke" recruits is not having polished boots. Also, make sure your pants are tucked in neatly and tightly into your boots. Constantly ask your friend (battle buddy) to check if you are wearing your uniform properly.
51. **Punishment is good for you.** During the first few weeks, the sergeants love to "punish" (they call it motivation) recruits. Instructors will even make up odd reasons to have recruits do push-ups, or leg lifts. If you find yourself doing excessive amounts of punishment exercises, it is normal. The purpose of these frequent punishments is twofold. One reason is to emphasize the importance of following orders. The other reason is to get recruits in top physical shape.
52. **Always respect the American flag.** During basic training you will learn how to care for and salute the American flag. Anytime you hear the National Anthem (or other designated patriotic music), immediately STOP what you are doing and stand at the position of attention. Always salute the flag when required. Remember, countless people have died for the flag and it is the most sacred symbol at your training base.
53. **You *can* finish basic.** You should keep in mind that the overwhelming majority of recruits make it through basic training. Basic is tough, but certainly not impossible.

Millions of people have completed training and so can you. Make the best of your experiences, be positive and remain focused. You will be successful with the right attitude.

Chapter 5

Reception Battalion “In-processing” and Your First Days at Basic Training

The day you are scheduled to leave for basic training, your recruiter will usually drive you to the Military Entrance Processing Station (MEPS) for the last time. From MEPS you will be taken to a transportation center such as an airport, train terminal, or bus station. Let’s say you are taken to an airport. You will board a plane with other recruits to travel to your basic training site. Thousands of new recruits are traveling to the same location everyday to start training. You could arrive at the training base at 3:00 in the morning and only get one hour of sleep (like I did). From the second you enter the military base, sleep deprivation and fatigue are your constant companions. As you wake up at 4:00 a.m. by roaring drill sergeants, you realize this is not home. However, you are not at basic training yet either. You are at reception battalion for “in-processing.” The next few days, weeks, and months will be new, strange, and sometimes stressful to you.

The subsequent chapters attempt to give you as much information about what to expect during your training as possible. Most insider tips are straightforward and objective. On the other hand, other tips are subjective and based on my opinion. However, the more that I can explain some of the seemingly crazy methods of boot camp, the less anxiety you will experience. Again, the more knowledge you have to de-mystify training procedures, the less stressful basic will be for you. Remember, the biggest factor for success is your attitude.

What to Bring to Training.

Simply stated, bring as little as possible to basic training. I laughed at new recruits who actually brought suitcases full of personal items and clothes. Basic training is NOT like summer camp. The less you bring, the better you are going to be when you travel home. The military will issue you a huge amount of gear to haul around. Your TA-50 duffel bag will contain at least two sets of everything—boots, uniforms, underwear, etc. Your very large TA-50 duffel bag will weigh about 50 pounds and will hardly close. The only nice aspect of the TA-50 is that all the gear is issued free to you.

When I traveled to Ft. Benning, I carried a very small (12 inch) gym bag which had one change of clothes, my toothbrush, shaving material, foot care products, boot polish and about $50.00. The previous items were all I brought and I was glad it was not much. The bottom line, pack very light and only bring essential items.

A List of What to Bring

1. A 9 x 12-inch clasp envelop to keep your important documents. (The envelope is optional, but it helps to keep papers, such as your official orders, organized).
2. Toiletry items: Toiletry kit, toothbrush, toothpaste, small comb, small soap, shaving gear, and anti-perspirant. (I suggest *anti-perspirant*, not deodorant which only covers up odors). Do not overload your kit.
3. Some people might tell you to bring two changes of civilian clothes, but you can get by on one pair. Most likely, you will be issued your military dress uniform (BDU) immediately the next day. You will not wear any civilian clothes or jewelry after you arrive at basic.

4. Bring two pair of plain white athletic socks with no markings on them. I suggest buying a high quality "Cool-max"+ cotton blend sock. 50% Cool-max socks are good.
5. A change of underwear, but more will be issued to you.
6. Bring, or wear while traveling to your military base, a good pair of running shoes. As mentioned in Chapter 3, you do not need to spend more than $70.00 on a good pair. Mid-level Nikes, Asics, or New Balances are decent choices. Make sure you do NOT buy cross-trainers, basketball, or tennis shoes—your shoes must be made specifically for running.
7. A quality pen that will not leak ink and a small notebook. You will take notes and write home often. Bring a few stamps and mailing envelopes.
8. A waterproof personal bag. Some people buy freezer zip-lock bags which are fine, but I suggest buying a high quality waterproof document holder. I bought my waterproof holder at the base's PX.
9. Bring a high-quality pair of <u>thin</u> athletic shoe insoles. Make sure the insoles are not too thick because they can squeeze your toes. You will want to change out your combat boot insoles because the boots come with insoles which are as supportive as wet sawdust. I suggest you try Dr. Scholl's athletic shoe, or work boot insoles.
10. Bring extra black paste wax *and* liquid boot polish.
11. Bring a small polishing cloth for your boots.
12. Bring extra pipe cleaners and cotton balls to clean your weapon. Not all basic training cycles will allow these previous items, but it was not an issue during my training. We spent hours constantly cleaning our weapons and consequently used a large amount of cleaning supplies.
13. Small personal towel.
14. Bring a limited amount of cash, or a credit card. Usually $50-60 is more than adequate. You will be issued money from

your first paycheck to spend. If you bring a charge card, I suggest bringing a credit card instead of a debit card. If your debit card is stolen, it is more difficult to recoup your money than it is for a credit card.

15. (Optional) You can buy shower slippers at the PX, or you can bring your own pair. Shower shoes are good to wear in the shower to prevent picking up a nasty foot fungus. You can also walk around the barracks in these flip-flops. The pair I purchased at the PX were very thin and cheap, but they worked fine.
16. (Optional) You cannot wear any type of jewelry during basic training. However, some drill instructors will allow you to wear a watch (my platoon was strictly forbidden from wearing watches.) If you are allowed to wear a timepiece, make sure it is very rugged. Casio makes "G-shock" watches which are a good choice.
17. (Optional) You can buy a pre-paid phone card, or you can purchase one at your PX. Using the phone at basic is a rare event. I bought a 30-minute phone card and had about 20 minutes left on it when I finished. Wal-mart sells cheap cards.
18. Two small, but strong locks with keys. Do not bring combination type locks--these locks are usually not allowed and you might forget the combination.

Important Documents to Bring to Basic

I purposely separated this section from the above because the documents you need to bring with you are very important. As mentioned, keep your documents neat and in a large envelope which is clearly labeled with your full name and phone number.

1. Bring your official training orders. (Make back-up copies).
2. Bring a *copy* of your enlistment contract.

3. Marriage, or divorce papers.
4. A voided bank check from your active bank account. Form SF 1199A is necessary to have to deposit money into your account. (See your recruiter about details).
5. A social security card, birth certificate and driver's license if you have one.
6. Addresses and phone numbers of friends and family written on one page of paper.

Special note:

I know this has been stated, but losing personal items to theft is commonplace during training. Carefully lock up your items, especially your personal documents. Bring copies of your documents whenever possible instead of the originals.

Another suggestion is that your local Wal-Mart or similar discount retailer can supply you with the items you need to take with you. You do not have to spend much money, but try to buy high quality durable materials.

Do Not Bring the Following Items to Training

Shortly, I will fully describe the shakedown process, but be advised that many recruits have had personal items taken from them permanently by sergeants. For example, a friend of mine brought a $30.00 small Leatherman knife that he had to surrender to the night sergeant. I was amazed by how many people brought tobacco, small radios and electronic games. During your processing at reception battalion, you will be told what you cannot have in your personal bag. You are given only one "free" chance to give up items that are not allowed. If you are caught with contraband after the amnesty period, pray for mercy.

Do Not Bring:

1. Jewelry, charms, bracelets, chains, necklaces, or the like. I would suggest you not wear your wedding band, or ring—leave it at home in a safer place.
2. Anything electronic such as radios, computers, gameboys, or flashlights.
3. Magazines, books, and no pornography.
4. Tobacco products, alcohol, over the counter drugs like aspirin, candy, or gum.
5. Prescription medications must be cleared by your drill instructor, or military physician. Additionally, do not bring knee wraps, compression shorts, or ankle braces. Only military doctors can approve anything medically related.
6. Anything that can be used as a weapon such as mace, pocketknives, or batons.
7. Boxer shorts, wild underwear, pajamas, robes, or fancy clothing.
8. Perfumes, cologne, hair gel. The only thing you bring with you that smells nice is your anti-perspirant and bar soap.
9. Cosmetics such as lipstick and nail polish. Do not bring any skin care product other than body soap.

Common Sense, but Needs to be Stated

Some people have no concept of the rigors of basic training. Although the vast majority of readers have some understanding regarding the extreme restrictions during training, the following must be stated as a courtesy to recruits who may be from remote areas, or have misinformation (especially from those silly reality shows on basic training). During basic training, you substantially give up your civilian freedoms to undergo very intense military training. You do not watch TV, use the phone, drink sodas, have free time, drive off base, read newspapers, or play games. You cannot have family visits, or even speak to your family unless given permission. The only communication you will have with the outside world is through the mail. Your life will be under the total control of your drill sergeants. Toward the end of your training cycle, you might get a few privileges such as using the phone for five minutes, but the first few weeks of basic are extraordinarily hectic and certainly stressful.

Life at the Reception Station

Reception battalion, or reception station, is a formal holding area for all new recruits arriving on the military base. Reception's purpose is to carefully process official military records and documents for all trainees. Recruits must be issued the proper gear, have medical / dental examinations, receive financial counseling and learn basic military procedures before actual training can take place. The following is just an example of a typical stay at a reception site. Your experience could be different, but not by much.

The term "reception" can be misleading in that the civilian form of word implies something positive. From the moment you are escorted from your point of arrival to the base, you will know you are not a civilian anymore. You will notice that no one is smiling, laughing, or acting casual. In fact, what you normally hear are people yelling at new arriving recruits. The yelling and barking of orders is unnerving to say the least and will be the most difficult aspect for you to become acclimated. Immediately, upon entering the base, the reception sergeants assert control over you and take away your freedom. From this point onward, you are told what to do and when to do it. You will definitely feel the intense climate of the station and you will be emotionally stressed. My first piece of advice is to try to understand that the strict environment and yelling is just part of the "game." Remain calm, focused and notice your reactions. You must give yourself time to adjust to this environment.

Sergeants run the show at reception battalion and are unofficially called "Smokies" due to their distinctive drill sergeant hats. When your bus/van arrives at the base, a sergeant will escort you to a large processing room with several other recently arriving recruits. You will be given a mountain of forms and documents to complete. Do <u>not</u> start writing and filling out the forms until after the sergeant says, "start writing." I remember many nervous recruits starting to write on their forms before permission was given and they were quickly punished. Even though the papers you will initially complete are easy and straightforward, double-check your work for errors. Again, you will be fatigued and stressed, so focus on the task at hand and listen to the sergeant's instructions very carefully.

Next, your group will be briefly educated about the "amnesty" room and "moment of truth." Besides telling you about the things you cannot have on the military base, the sergeant will elaborate on the consequences of possessing contraband and not disclosing important information. In short, if you have items which you should not have, you will be living in a world of pain. Fortunately, the sergeants allow you to enter a small empty room to give up those items on the contraband list. Even if you bring a switchblade knife, you will not be in trouble if you give the knife up in the amnesty room. Additionally, you will be given an opportunity to disclose personal information which you have concealed up to this point. This "moment of truth" process is given only once and is brief, so take advantage of it.

After your paper work is complete, you will be assigned a roster number, a bunk bed and to a group of people called a squad, or platoon. Again, it is extremely easy to forget important things when you are stressed. Focus and memorize these previous assignments and numbers.

If it is night, you will be ushered to your assigned bunk bed and locker. Once at your bunk, put away your personal items in a "locked" locker and go to sleep quickly. The beds are about as comfortable as camping cots and have a distinct musty smell. Do not be surprised if you only receive one or two hours of sleep the first day. You will learn very fast that sleep is one of the tools sergeants use to control recruits. Start conserving both your physical and emotional energy from the first day.

The next morning about 5:00 a.m. the sergeants wake your barracks by turning on the lights. Although the sergeants speak in a loud voice to signal wake-up, they do not startle everyone by yelling. (I believe the training authorities have seen the research which says that people just waking up are not fully rational due to a lack of blood flow to parts of the brain.) However, do not be too surprised if you are suddenly and rudely awakened during actual training. As you slowly come out of the fog of sleep, you see that your barrack's room is quite large, maybe 50-60 recruits. Trainees rapidly start moving toward the showers and bathrooms. Suddenly, you realize that another freedom has disappeared—your personal space. For the next few months, you are constantly within arms reach of someone else, even in the bathrooms.

The sergeants re-enter the barracks and tell you how much time you have left to report outside in formation. Generally, you will have about 15-20 minutes to wake, shower, dress and make your bed. As you report outside, you will be placed in a formation of about 40 trainees. This basic formation is your reception battalion platoon. It is sometimes beneficial if you are placed near, but not in the middle of this formation. As with your newly assigned identity number, remember your platoon's name and your place within it.

In a large, but confined area, you will notice other platoons and sergeants yelling out instructions. It is still very early in the morning and the yelling starts to bring back the stress which left you as you slept. After roll call, in which you must answer with a loud and snappy, "here," you will be marched to the cafeteria (defac, or chow hall). Your platoon, as well as many others, will be waiting outside the cafeteria because hundreds of other recruits are inside already eating. While standing in line, you will experience the "hurry up and wait" nature of the military. When the sergeant receives the clearance to allow your platoon to enter the chow hall, you will have to scream out your last name and assigned number (or at least count off). At every meal, there is usually an accounting of all who enter and eat in the cafeteria.

As you enter the cafeteria, remember to remove your cap, or headgear. You cannot wear your cap inside of *any* building. Sergeants love to catch new recruits wearing their headgear inappropriately. If caught, you might be doing push-ups with your face in your oatmeal bowl. My first meal in the chow hall was not too pleasant because the drill sergeants constantly roamed the aisles looking for unsuspecting screw-ups. Bunched up tightly, your platoon quickly picks up trays of food and utensils. Generally, you cannot ask for certain food items; you just take whichever tray is available. Some trays may have different foods than others, but all are balanced meals. Do not be surprised if the person next to you gets bacon and eggs, and you get pancakes and sausages.

Surprisingly, the military does not have bad-tasting food. The food quality is on par with a public school cafeteria, airline, or hospital. Although not the same as nice restaurant meals, military food is certainly healthier for you than greasy fast food which causes naptime. Unfortunately, if you do not like something you might still have to eat it. Before any recruit in my platoon could ever leave the cafeteria, we had to show our plates to the sergeant. If the food on our plates was not fully consumed, we did push-ups. Additionally, you have to eat very quickly. Some people say you have about 15 minutes to eat, but usually recruits are pressured to leave within eight minutes.

After leaving the chow hall, check in with your platoon leader to see when the next formation will be held. Sometimes you might be able to go back into your barracks to clean up, or even rest. Remember, this is reception battalion, not your actual basic training area. Therefore, you will have many more pockets of free time until you ship out "down range" to start training in earnest. My advice is to take advantage of any rest time you can get at reception.

Normally, you will only have about 10-15 minutes after eating before the next formation. During your next platoon formation, your reception sergeant marches you to the supply building. This huge building is where you are issued all of your training gear. Your gear is free, but you are responsible for every item given to you. As with the chow hall, you must wait in a large group before entering the supply building. When your group enters the issuance area, you are given a huge green canvas duffel bag. In this duffel bag, you will be given various items which comprise what is called your TA-50. First you are measured and then given a camouflaged uniform called the battle dress uniform (BDU). The BDU is a sturdy, durable uniform you wear everyday. All soldiers receive four uniforms—two summer sets, and two winter sets. Your last name is embroidered on the front of your BDUs.

Other items which you rapidly cram into your huge duffel bag are two sets of boots, many sets of underwear, military issued socks, physical training (PT) uniforms, winter jackets, warm-ups, towels, handkerchiefs and loads of other gear.

After you are issued your TA-50, you are marched to a small store called a Post Exchange (PX). Some stores are bigger than others. Do not be dismayed if your PX is small because it will have all the necessary supplies. Your sergeant will tell you what you will need to buy, but he/she will only give you a few minutes in the store. Also, you will not visit the PX many times (maybe twice the entire time at boot camp) so stock up on needed items. Common items which need to be purchased are bar soap, boot polish, toothpaste, running shoes, batteries and deodorant.

At this point, the reception sergeant allows you to return to your barracks to secure your newly issued gear into your footlocker. It is of the utmost importance that you lock your gear because stealing is common. After organizing your area, it will probably be lunchtime. Again, you will stand in formation outside the chow hall and wait for your platoon's orders to enter. Example of lunches are hamburgers, chicken sandwiches, and salads.

Even though you may not be hungry, you must go to the cafeteria and consume food. After a few days, you might start dreading the chow hall because you have to eat when you do not want. The reason behind encouraging new recruits to eat three square meals per day is twofold. First, there are hundreds of recruits who have had unhealthy eating habits for most of their lives. Many recruits might even be malnourished. Therefore, before actual training can start, all recruits must have roughly the same baseline nutrition. Second, the calories recruits burn when training begins is amazing. You could say the reception station is trying to "fatten" recruits before the real burn begins.

Ironically, I was one of those recruits who hated to eat at reception battalion. The initial stress and shock of the military took its toll on my digestive system at first. After all, who really wants to eat if you've been constipated for three days. I even tried to give away food to my buddies during my stay at reception. However, when actual basic training started, I could hardly wait to stuff myself in the chow hall. When training begins, recruits burn off everything they eat and more.

After lunch, you will march over to the barbershop with your platoon. You silently wait in a long line to have your hair shaved off (for males). Ironically, you will pay about $8-10 to have the barbers shave your head like a farmer shears a lamb. Women usually just have their hair cut short. The shaving experience is fast and somewhat odd. You will be ordered to sit in the next available barber chair. The barbers by mid-morning seem to operate like robots, monotonously shaving hundreds of people's head the same way. The barber will grab your head firmly and turn on his commercial grade clippers which sound like a bag of angry bees. Sometimes a barber will go somewhat slow and work the shears around your head. Unfortunately, my barber moved my head wildly against his stationary clippers. Few recruits leave the barber chair nick-free. Do not be surprised if your completely bald head is covered with tiny red nicks.

Having a clean-shaven head is the first prominent visible sign of being in the military. It will feel very weird, but the reason the military makes you bald is not just to aggravate you. A bald head keeps the body extraordinarily cool in the warmer months. Also, a bald head is much easier to maintain. Finally, and primarily, the military is making everyone equal by taking away individuality. The military shaves your head to vividly demonstrate its authority over you and to make all recruits the same. Everyone starts training as a "new" person.

After recovering from your buzz cut, you will find yourself marching over to a building to fill out more paperwork. You will be issued official identification cards and metal tags with personal information (dog tags). You will take another medical and dental examination, get X-rays, and immunization shots. The shots you are given are administered with bulky hand held "injection guns" designed to quickly immunize scores of people. The shots do not hurt much, but there is a noticeable sting. You will get a shot in both arms and have to hold a cotton ball for several minutes over the area. You might be issued military glasses if your eyes need corrected vision (these standard issue glasses are so heavy and ugly that they are called "birth-control" glasses).

After dinner, which is always served promptly at reception battalion, you may find yourself with some down time. Or, your platoon could have a class on common military tasks such as making your bunks properly, or how to march. Also, the sergeants might have you doing physical exercises to prepare you for later training.

The next few days the reception sergeants will have you fill out more papers, have many different people speak with you regarding military topics and keep you busy with physical conditioning. After about three days at reception, you are just waiting to be shipped down range to start basic training. Sometimes, as in my case, you could spend longer than a week at reception station waiting for paperwork to process and be assigned to a training company.

If you do spend longer than three days at reception, the sergeants there will try to keep you busy. Most of the time, physical conditioning exercises are a favorite activity of the drill instructors. In addition to learning physical training (PT), formations and drills, you will also learn valuable tasks which will be useful when real training starts. Use your time wisely at reception and study your IET book.

Additional Notes About Reception

Rest assured, reception battalion is very stressful at first. The sergeants there will constantly yell at recruits and even "smoke" them (make them do push-ups). Punishment exercises are frequent. You may be formally taught elementary tasks like standing in formation, marching, polishing boots and how to do PT. However, the true purpose of reception is not training, but rather the processing of official personnel records, issuance of gear and assignment to a training company down range. After just a few days, you might see the attention shift from your platoon to a newer platoon which needs records and training gear. The stress you experience is due to the stark contrast between the civilian world and the military. The reception sergeants want to purposely emphasize that recruits are not at home anymore.

Most of the time, reception sergeants unofficially assign platoon leaders. These leaders are usually recruits with ROTC (officer training) experience, higher rank, or college education. Even though these leaders will not necessarily be your platoon leaders when basic starts, they will serve as the line of communication between the sergeants and the platoon. I do not advise being a platoon leader at reception because they do not get paid any extra money, or have privileges. Most of the time, our platoon leader was busy doing push-ups because he was blamed for other recruits' mistakes.

Another special note concerns physical training (PT) at reception. Recruits are introduced to how PT is correctly performed in the military. You will attend morning PT formations and exercises. You might even take a practice PT test. Usually, out-of-shape recruits are discovered by this test. If you perform poorly on the reception's diagnostic exercises you will be place into a Fitness Training Unit (FTU). The FTU recruits are known unglamorously as "fat boys" and they constantly do extremely tiring workouts. These special units keep their recruits until they can meet minimum standards. It is wise to be in decent shape before basic training because the FTU is highly unpleasant.

The last suggestion I have is always make good use of your time. If you have any free time, use it to regain your energy by properly resting, or reading your IET book. As mentioned, some basic tasks might be taught during your time at reception. Therefore, if you learn how to correctly polish your boots, or make a bunk, learn it well. Once down range, learning tasks will not be as easy.

Leaving Reception Battalion In Style—Cattle Cars

When you first arrived at your reception site, the new environment was strange and very nerve-racking. However, after just a few days, you became used to the strict procedures and structure. All of your records have been processed and you have been issued a king-sized duffel bag full of gear. The reception sergeants have turned their attention away from you to focus on all the other constantly arriving recruits who still need to be in-processed. Actually, by the fourth day, things seem like they are winding down and not so bad after all. Regrettably, events are going to take an unbelievable turn.

On the fourth or fifth day, your reception platoon will be instructed to prepare for departure "down range." You will sit outside in formation, dressed in your BDU's and with all of your personal gear. Sitting atop your sofa-sized duffel bag, you will notice several eighteen-wheeler trucks parked outside. These special trucks are towing huge metal trailers used to haul cattle. Suddenly, a sergeant screams your roster number and you run lugging all your heavy gear to the cattle trailer. As you struggle to get your duffel bag aboard the trailer, you are literally squeezed into the other 40 recruits. Indeed, you feel like an animal getting ready to go to the slaughterhouse.

The symbolism and psychology behind this process is evident. Before actual basic training begins, recruits are not worthy of respect and are treated as lowly cattle. From the drill sergeants' perspective, recruits have not done anything to earn status. Recruits are not seen as individuals, but as a herd. In a way, this psychological game is a mild form of dehumanization—which has a distinct training purpose. It could also be argued that this game is an illusion, in that it is orchestrated to make recruits "think" differently about themselves. In other words, the cattle car mind trick is the first real step toward stripping away your civilian attitude so you will conform to the total control of your drill instructors.

Your cattle car, full of squeezed recruits, will depart with about four other big rigs. You will drive around the base, sometimes doubling back to confuse you as to your precise location. The military does not want you to know exactly where you are---which purposely creates more anxiety. About 10 minutes after you have boarded the sweltering hot trailer, the truck comes to an abrupt halt and the doors scream open. Read Chapter 7 to find out what happens next!

Chapter Summary

1. You will be transported at government expense to your basic training site. You, and thousands of other new trainees, will converge at your assigned military base. The massive influx of recruits must first be "in-processed" at a specific place on the base called reception battalion. Reception is not the place where your hard-core training starts. Instead, in-processing entails new recruits receiving, medical examinations, financial counseling, haircuts and all required gear.
2. Recruits arrive at the reception site during all hours of the day. Do not be surprised if you only have two hours of sleep the first day. You will be stressed the first few days at reception, but by the third day you should start to adjust. Due to your elevated emotional state, it is imperative that you remain relaxed, but very focused on what is said to you. Remember your assigned roster number.
3. Even at reception battalion, do not stand out. Be quiet and do what you are told without delay. Make good use of any free time to catch up on some much needed rest. Study your IET book during breaks.
4. Do not over-pack for basic training. The lighter you pack the better, but make sure to bring essential items. Some im-

portant items include all official documents, toiletries, a change of clothes, and running shoes.

5. Do not bring illegal (contraband items). Review the "not to take" items. You will be given a chance to give up items you should not have packed. You will also have a chance to tell the truth about things you concealed. These amnesty and moment of truth periods are only given once, so take advantage of them.

6. My example of reception battalion life may not be exactly like your experience. However, my example should give you a general feel for some activities you might encounter and the purpose of reception.

7. Lock up all of your personal items in a secured area. There is a good chance that someone in your platoon will be a victim of theft, so watch your gear.

8. Do not bring too much cash to basic training. You will be given an advance from your first month's paycheck to buy additional items you need. Manage your money wisely; some recruits buy unnecessary things and waste their money. If you bring a charge card, bring either a credit card or pre-paid card. Unfortunately, these items are targeted for theft.

9. Have a positive attitude and realize that basic training is about mind games. The drill instructors want you to "think" the worse will happen to you, but usually things will not be too difficult. To get you through tough times, focus on your goals.

Chapter 6

Mind Games of Basic Training

Before I start explaining your basic training life once you leave reception battalion, it is important that you prepare mentally. I am positive that the next two chapters will not make me popular with recruiters, or drill sergeants. Because I am giving away the sergeants' secrets, I have noticed attempts by some people to discredit this book by writing criticisms under fictitious names. Regardless, if I can make the reader more focused and less stressed during training, I will have accomplished my mission.

The purpose of this section is to give a recruit an idea of the "games" the military utilizes during basic and to explain the structure of training. By understanding the general structure of training, you will have taken away some of the drill sergeants' power over you. Next, I will try to elaborate the logic behind some of the strange and stressful tasks you will encounter. Also, I will describe some of the techniques the military employs to turn civilians into soldiers. Granted, most of this section is based on my opinion, but you will see my rationale is intuitive.

The Phases and Psychology of Basic Training

There are three broad, but distinct phases to basic training. Each phase lasts about three to four weeks. Sometimes, completion of one phase and entry into another phase is not marked by any type of ceremony. You might not even know when you are in a different phase. However, some training posts will have a flag changing event in which your platoon, switches small flags on the company's flagpole (guidon). The phases may have names such as the Courage Phase, or Patriot Phase. Again, each training base has slightly different styles for implementing training procedures.

The Phases are:

1. **Total Control Phase (red flag).** This phase is extreme and most restrictive. Everything you do will be strictly controlled. Orders are sternly yelled and punishment exercises are a very common occurrence.
2. **Phase 2 (white flag).** I call this stage the "mixed" phase because drill sergeants ease up slightly on total control and start adding more meaningful tasks to your schedule. However, some days during Phase 2 can feel like Phase 1.
3. **Phase 3 (blue flag).** In this phase you will feel more confidence as the drill sergeants treat you with a little more respect and you are given more responsibility. Your tasks during this phase will focus on self-reliance and confidence building. Although you will have more personal time, your drill instructors still watch you closely and can dish out punishments quickly.

Psychology and Training Techniques

From the moment you set foot on your training site, a drill instructor's primary job is to strip you of your civilian habits. I have been admonished for saying this, but basic training does not really teach you many significant skills; rather the focus is on *indoctrinating* you into the military lifestyle. Because the military life is so radically different than the civilian world, radical and stern measures must be used to change your behavior. It will help your stress level if you remember there is usually a reason for seemingly crazy military practices. Your stress level should decrease if you can figure out the meaning behind what the instructors are trying to do—even when they are barking obscenities at you. If you cannot think of a rational reason for what appears to be irrational behavior, you need to fall back on your positive attitude and just "go" with it. Remember, your thoughts and attitude can be your worse enemies if you do not have them tuned to a positive frequency. The following items are psychological techniques the drill instructors use during basic training.

1. Information Control and Fear

If a drill sergeant tells you to do 30 push-ups after you have volunteered to help carry a heavy water jug, you might think his orders are bizarre and unfair. However, trainees are purposely kept from understanding the reasons behind basic training practices and odd drill instructor behavior. By controlling information and what recruits know, trainees are thrusted into an "unknown" state of mind. The fear of the unknown is perhaps the foundation of most forms of anxiety. Indeed, much of human behavior can be described as people attempting to lessen anxiety by controlling the unknown future. Not fully understanding what will happen, or not understanding why something is happing fosters extreme stress and confusion in recruits. Once in this fearful state, a person has little choice other than to follow orders growled by sergeants. Recruits quickly associate people who yell out instructions with those who are in charge. Trust me, during the first three weeks of basic, the drill sergeants want you to know they are totally in charge.

The reason drill instructors want you to know that they absolutely control your life is twofold. First, there are only a few instructors per training company of 200 people. Therefore, recruits must do all tasks efficiently. There is no time to explain why activities are done a particular way. Second, obeying orders quickly and without question is the pillar of military discipline. A country cannot have an effective fighting force if lower ranking personnel do not follow orders without hesitation. Your drill instructor's mission is make recruits realize rank and orders *must* be respected and quickly followed.

2. Sleep Deprivation

From the time you arrive at your basic training site until the day you leave, you will be very tired. Some days you will receive only one, or two hours of continuous sleep. In addition, the constant stress of the environment will intensify your fatigue. Sometimes you will have to drag yourself out of bed at 3:00 a.m. to stand guard over a pile of sand. As you stand in full combat gear guarding a sand pile, you will wonder why the drill instructors keep you in this constantly tired state.

Fatigue is a tool that helps to weaken recruits' mental condition. As a group, people are easier to control when they are weary. However, if recruits are completely exhausted, it is more likely they will become frustrated. Drill sergeants are trying to find out how well you can handle frustration and other related negative emotions. Sergeants also understand that fatigue helps to break teamwork apart and they want to see the strength of your team spirit. Last, by creating a perpetual exhaustive climate, the drill sergeants simulate a combat experience. An effective fighting force must be able to operate under stressful conditions for long periods of time.

My suggestion is keep your team together by realizing the drill sergeants' purpose. Also, never show your true emotions on your face or body posture during training.

3. Purposefully Creating Frustration.

I personally believe that this is the favorite mind game that the drill sergeants play. Drill instructors will purposefully make your platoon do things to create frustration and other negative emotions within you. Sergeants will also make comments to you in an attempt to make you express emotions. For example, I remember one day my group was led to believe that the sergeants were easing up. We were on a two-mile run when the drill sergeants brought us to an area full of sand and mud. One sergeant stopped our group and made up an excuse to punish us—he said we were not running with spirit, whatever that means. Because we did not run with excitement, the sergeant made us roll in the sand and mud. Next, our group ran back to our barracks and was ordered to change clothes. Mysteriously, we did not change fast enough and we had to re-don our PT uniform for more exercises. As a final rub, another drill sergeant reported that our barracks were full of mud (which could not have been helped due to rolling in mud) and consequently we were taken for another dive in the pit.

The reason sergeants try to create stress is to see if you can handle your emotions. Drill instructors must instill a "military bearing" in their recruits. In fact, you will hear several times never to lose your military bearing. A military bearing is the ability to calmly control your reactions to events, even traumatic events. Again, the rationale is that during a combat situation, being able to handle stress is advantageous to survival. Unfortunately, the military does not teach recruits effective stress reducing techniques.

I suggest that you never smile, smirk or grimace to a drill sergeant. Learn how to stare blankly with eyes fixed straight ahead. Even when a drill sergeant is yelling two inches from your nose, do not blink, turn away, or show your discomfort. A helpful tip is to visualize in your mind's eye that you are a brick wall and all of the drill sergeant's words are harmlessly bouncing off of you. Do not take anything said to you personally. Remember, when you control your emotions, you are able to think clearly.

4. Hard and Easy Contrast

For the first part of basic training, you will do nothing right. Even though the drill sergeant could not do an assigned task better than you did, your only reward is a barrage of insults. Even simple tasks will not be up to the irrational expectations of the sergeant. To be fair, sometimes if you complete a task well, the drill sergeant may not yell at you, but the only reward you get is no response at all from the instructor. It is best that you do not expect any rewards or praise until the last phase of training. Resolve yourself to the fact that nothing you do will be up the sergeant's standard until Phase 3.

The primary cause of making simple tasks seem extremely difficult is to create an illusion in the minds of the recruit. If a task is too easy, then people do not assign value to it. Things which are difficult are seen as significant and worthwhile. After all, if any- body can easily do something, it is not special and has little perceived value. The drill sergeants want to make you believe you are doing something hard because it instills a strong sense of confidence and worth. For example, the task of making a tight bed is constantly met with your bed being turned upside down by the drill sergeant. Then in the last phase of training, the sergeant exclaims, "Now that's a tight military bunk. Good job soldier!" By being hard on you initially, then slowly letting up, you feel as if meaningful progress is "earned." Additionally, the praise and compliments drill sergeants give are perceived as sincere—therefore the praise is seen as more rewarding.

Another side effect of starting training under the illusion of it being hard is that it helps recruits pay attention to details and instills discipline. If you know the drill sergeant is going to drop you for 40 flutter kicks for not having your boots shiny, you are probably going to pay attention to the details of polishing your boots.

5. Personal Put Downs and De-individualizing

Writing about this aspect of training is controversial. Some people or sources (TV shows) may have you believe that cursing, or rough physical handling is not a common occurrence during training. In my opinion, the previous events are a regular part of basic. In fact, I never heard so many colorful ways to use the words, "shit," "asshole," or "fuck" in my life. To say drill sergeants do not curse at recruits is like saying men do not think about sex.

Although drill sergeants are professionals, their use of strongly phrased words and denigrations should be viewed as a training tool, not as mere insults. First, when a drill instructor is personally screaming at a recruit, he is probing the recruit for an emotional weakness. Again, sergeants try to push the emotional buttons of recruits to assess the strength of the recruit's military bearing. In other words, the true measure of toughness is how much self-control you have within yourself. Next, personal attacks are designed not only to emphasize who is in charge, but also to gauge how much you will respect the omnipotent chain of command. Recruits must always respect orders and rank, or the military will not function effectively.

Drill instructors during the first phase try to de-individualize you through the use of intimidation. In simpler terms, the drill sergeant does not want you to think about yourself, but about your team. If you can think about your team's goal and remain focused when the drill sergeant is yelling at you, then you will have achieved an important military goal. Individuals in the military are required to sacrifice themselves for the greater good of the group. Additionally, drill sergeants degrade trainees because they do not want recruits to think of themselves as worthy until they "earn" respect. The harder a person works at an activity, the more he/she will feel the activity is worthwhile once completed.

Side Note: While doing recent research for this guide, I noted that some instructors act differently when they know they are being observed or videotaped. In psychology, we call this the Hawthorne, or Halo effect. Therefore, if you watch a documentary about boot camp, you will <u>not</u> see drill sergeants call a recruit an "asshole." It is human nature to want to portray a positive image when we are officially observed. However, I can attest from actual experience that drill sergeants can say and do some extremely offensive things. I do not make the previous statement to frighten new trainees, but just to state a blunt fact which other sources do not clearly state. Usually rougher basic trainings are directly related to single gender combat units. In my opinion, if a person cannot take constant cursing and some tough handling, the military will not fit his/her interests.

5. Good Sergeant, Bad Sergeant.

Many movies have a "good cop, bad cop" scene. The bad cop harshly interrogates a suspect and gets the suspect nervous. Then, when the bad cop leaves the room, a good cop comes in and gets the suspect to open up. While not all platoons will have a good (nice) drill sergeant and mean drill sergeant, this combination is not uncommon. Sometimes, this combination is merely a reflection of the sergeants' natural personalities. If this drill sergeant structure exists in your platoon, it is important to realize how it is used. Sometimes the good (less feared) sergeant can assess the platoon's behavior from a different view, or see who is having difficulty adjusting to military life. The good drill sergeant usually reports back to the head drill sergeant about the overall attitude of the platoon and other things of interest. My suggestion is do not try to buddy up to the nicer drill sergeant. Always remain, serious, calm, and focused around any sergeant.

Another point of this section regards the drill sergeants' use of contrasting attitudes. One day your drill sergeant may be really cool and friendly. This attitude lulls you into thinking training is becoming easier. Then suddenly, the same drill sergeant does a schizophrenic about-face and goes into an irrational rampage of dishing out punishments. Although there are times when the drill sergeant will be genuinely angry, mostly it is just an act.

Because the drill instructors can vary their attitudes daily, it is important that you remain on an even emotional keel.

Insider's tip on dealing with the "mind games"

The best piece of advice I can dispense when it comes to mind games is to understand your response to other people's actions. How you react to situations will in turn influence the situation. Therefore, the best initial response to a drill instructor growling at you is to remain very calm, but focused. By remaining calm, you do not give your emotional brain a chance to hijack your "rational thinking" brain. I know the previous sentence may sound unusual, but there is actually a biological process at play when you express emotions. Additionally, even though a drill instructor may be hell-bent on putting fear in you by yelling in your ear, remember that no physical harm can come to you. Even in the most stressful situation imaginable, the one element of control you have is how you respond. You *always* have a choice—that of responding maturely, or immaturely.

It is also important to remember that drill sergeants are using techniques which are designed to look strange. By realizing that things in basic training are created to *appear* crazy and irrational, you should feel less stress because you know something that is supposed to be unknown. Humans have a need to assign purpose to things, especially behavior. The drill sergeants' game is to act odd to create stress because you won't be able to assign a rational purpose to what they do. The drill sergeants' behavior may seem strange, but this is an "act" to get you to think what they want you to think. Try to see past the mind games of the drill sergeants and figure out the reasons behind their behaviors and activities. Ask yourself if the sergeant's game is designed to control you, get you to show emotions, or motivate you. Simply stated, never take anything the drill sergeant says about you (or your mother) personally because it is a way to get you emotional.

Many people try to alleviate stress in their normal lives by predicting future events. Indeed, people *must have* a sense of the future because it gives them hope. Unfortunately, boot camp is meant to be unpredictable and it will naturally raise your anxiety. My advice is to remain future-oriented by *visualizing* yourself where you will be a year after boot camp. Where will you be six months from basic? Try to picture in your mind all of the minor details of how you will look and the things you will have after graduation. When times are extremely demanding, do not dwell on your present situation, but on a positive visual image of your personal future. Remember that things will get better in time. Adopt a "go with the flow" easygoing attitude. Remain serious, but do not get upset when the drill sergeants make you do irrational tasks—just go with it and think about next year.

Another method to help you deal with stress is to self-monitor your mental and physical status. Talk to yourself using your inner voice. Use self-talk to make yourself stronger by reinforcing a positive attitude. For example, on a painful 10-mile road march, assess the pain you are feeling. Then, tell yourself, "Whew! This march is hurting my legs, but I can make it. Even though it hurts, it's not going to stop me and I'm only getting stronger—much stronger from it!" Your inner voice should constantly echo,
"Nothing worthwhile is easy!" Do not underestimate the power of self-talk to build a positive attitude. You can overcome almost anything if you develop beliefs which do not have limits. Never tell yourself you cannot do something. If your beliefs begin to falter, remember this quote, "Whatever does not kill you, only makes you stronger."

Do not just limit yourself with inner speech. Some recruits talk with other recruits when their frustration level boils over. During those times you have away from the sergeants, talk with fellow trainees and develop friendships. Write positive letters home and re-read letters from friends. Talk with friends about humorous events of the day. Finding *humor* in stressful things is a valuable talent and will help you forge bonds with others. People need to connect with other people in order to be well-adjusted. Recruits who are loners do not do as well in basic as those who have a solid social network. It may not appear to be so, but drill sergeants fully understand the importance of recruits bonding with each other and will allow the platoon the necessary time to construct friendships.

The last way to deal with the mind games and stress of training is to remain focused on your goals. If you have a true purpose and meaning to your life, you can achieve great things. A potent suggestion I made earlier was to think about three significant long-term goals in your life. Ask yourself what you really want in life. What makes you excited? It is appropriate to attach an emotional tag to your goals—it will motivate you to reach your goals. Having meaningful (valued) goals in life is really the secret to life itself because it provides hope. Your current attitude and future will be founded upon your life's goals. Remaining *intensely* focused on your objectives will not only remind you of the true reasons why you are in basic training, but it will carry you through to the end.

Summary

1. There are three separate parts to basic training in the Army. Most likely, other branches of the armed forces will have distinct phases during their training. By realizing that your training has different phases with different purposes, your stress level should lower because you will know what to expect. It is most important to remember that the first phase, which is the harshest, only last about three weeks.

2. Realize that there are specific reasons behind seemingly irrational and nonsensical tasks during training. Some psychological games drill sergeants use are: controlling what you know, the use of fear, creating frustration, making things seem hard when they are not and changing their attitudes on a daily basis
3. Try to counter the psychological games used by drill instructors by understanding that many of their behaviors are "acts." Drill sergeants want to create the illusion that they are harsh, insensitive people. Actually, most drill instructors are normal people when off duty.
4. Other ways to deal with mind games and stress are to realize you always have control over how you respond to events. People always have a choice in how they value situations and react to them.
5. It is vital that you write letters home, make good friends, and stay focused on meaningful goals.
6. Maintain a positive attitude and realize that basic training, like all stressful situations, will end in just a few short weeks.

Chapter 7

Life in Basic Training: What to Expect

This chapter's intent is to give the reader a general idea of what to expect during each of the three phases of training. I have compiled information regarding basic training from notes as a recruit myself, interviews, research, and recent documentaries. Again, your experiences might be slightly different depending on your training site and your type of personality. As noted in previous chapters, basic training is hard, but not impossible. In general, you should follow three primary rules during the first part of basic training: <u>do not volunteer for anything; do not show any type of emotion; and do not stand out from the group</u>. Most people complete basic and you can also. At the end of just a few months, you will unquestionably feel stronger and more confident as you graduate boot camp.

The Arrival of the Cattle Cars

We left off in Chapter 5 with recruits being crammed into hot cattle cars at reception battalion. The following illustration is a mixture of my experiences and those of others in my training company.

When the big rig trucks finally stopped, drill sergeants opened the steel doors and literally yanked recruits out. The scene was chaos in the extreme. Picture hundreds of anxious trainees running in every direction with their large and heavy duffel bags. Sergeants, the size of WWF wrestlers, were getting in recruits' faces and yelling the harshest words devised by man. I remember a 6'3 sergeant running up to me as I reached down to pick up my personal bag that I dropped, "You goddamn stupid testicle head, what the hell is taking you so fucking long…move your goddamn ass or I'll kick it!" The sergeant was barking at me so loud and was so close to my face that the brim of his hat kept hitting my head. Moreover, I could see a vein on the sergeant's forehead nearing its bursting limit. Everywhere I looked, huge sergeants were acting insane with rage. To say recruits were frightened and confused would be an immense understatement.

As the sergeant wolves corralled the recruit sheep into a formation, we were instructed to leave our gear in a separate location. Next, guard dogs and men in black shirts combed through our gear to catch recruits who brought illegal substances. I cannot imagine what things happened to those recruits who tried to sneak in tobacco, or marijuana.

After our "shakedown," the drill sergeants began toying with the shaken recruits. I distinctly remember one drill sergeant bellowing out the command, "Okay, if you think your recruiter lied to you and you want out of here, take a step forward." As I fought to hold back the laughter of this obvious ploy, I knew some poor guy would not understand this trick. Sure enough, a terrified recruit stepped forward. The drill sergeants no sooner saw the trainee break formation than they surrounded him. The sergeants yelled at him, "Holy Shit! Looks like we have a flag burner! Get your ass in the back of the formation!" For the next several minutes, all we could hear were the grunts coming from a recruit who was being introduced to muscle failure push-ups and flutter kicks.

From our one large formation, the drill instructors called roster numbers. When they called your number, you had to quickly gather all your gear and run up to your assigned barracks. The sergeants assigned roughly 40 people to my platoon. A training "company" will have four platoons, or 200 people total.

Phase I: Total Control

The First Night at Basic

Do you think your first night at reception battalion will take some adjustment? Do you think the military will be stressful at the intake site? Well, prepare yourself because your first few hours down range at your basic training barracks will blow away any stressful time you have during training. I can imagine the drill sergeants the day before training relishing the arrival of their new recruits and how much fun it will be for them to harass trainees. Although I was nervous, I did manage to find humor in the drill sergeants' attempt to intimidate us. Actually, I was impressed by some of the novel punishment exercises and the fictitious reasons made up to discipline our platoon.

As our new platoon ran into our barracks (a large room with 40 bunk beds), we were ordered to put our feet on the long straight line in front of our bunks (called toe the line). Standing stiff at the position of attention and with eyes fixed forward, a group of three sergeants took turns screaming in our faces. There was even an assistant instructor who was a cadet at a military college who joined in the yell-fest. When one recruit barely blinked, the lead drill instructor made all the recruits lay on their backs and crawl backwards on the floor—even underneath the beds. We literally were cleaning the floor with our backs and moved around the entire room until we got sore. When a whistle blew, we had to immediately stand up. Of course, our platoon did not stand quick enough and we started doing push ups until our arms shook from exhaustion. This stressful routine went on for a few hours. However, the purpose of the insanity was to make us fear the drill sergeants enough so they would have "total control." Establishing total control is the primary mission of the first week.

The Shower

The extreme yelling and punishment exercises during the first few hours of basic were one way of showing recruits who was boss. The next means of intimidation came in the form of controlling how we bathed. The hot environment of Ft. Benning, Georgia in the summer made for some sweaty and smelly workouts. However, instead of taking a real civilian shower to wash away the day's grime, the sergeants had something else in mind. Stripped down to our underwear and holding a towel, our platoon formed a long line before the shower entrance. A drill sergeant with a stopwatch screamed that we had only 20 seconds per shower stall. As the whistle blew, a recruit would run in the first stall to find the showerhead barely dripping a few drops of water. When the whistle blew again, the trainee ran to the next stall to find another drip barely enough to get wet. This procedure went on four more times until the recruit left the shower to go to the sink area to brush his teeth (under supervision).

Even the smallest details of a recruit's life will be fully controlled. Most personal freedom is taken away during phase one. The purpose of controlling your freedom is so that you have to earn it back slowly. This total control process breaks down your civilian attitude and builds you into a tough military life.

First Few Days

The next day, you will be awakened at 4:00 a.m. The drill sergeant growls out the orders that your personal area must be ready for inspection *and* the entire barracks must be cleaned before first formation which is at 4:20 a.m. This is an impossible task and so you will have an extra hard morning exercise (PT) to motivate your platoon to do better. After about an hour of rigorous push-ups, sit-ups, flutter kicks and log rolling, you run back to the barracks to dress in your basic camouflage uniform called the battle dress uniform (BDU). The BDU set consists of heavyweight cotton long pants and a long sleeved shirt. The heat caused by wearing BDUs in Georgia's summer sun was indescribable. The salt produced by our heavy perspiration literally made rings on our clothes. This is why scheduling your training in the spring or early fall is best.

After recruits dress for the day, they fall back into their platoon's formation. The platoon is then marched to the cafeteria (chow, or mess hall). Around 6:00 a.m., you should be shoveling down a hearty breakfast. Although the food is above average and consists of bacon, sausage, pancakes, and eggs, it is difficult to enjoy. In the chow hall, sergeants are circling like hawks harassing recruits. If you talk, or sit with your feet not exactly right, you are targeted. Within first few days, it is not uncommon to see recruits doing push-ups in their trays of food. I remember a drill sergeant throwing a nervous recruit's tray across the floor just because he had an extra food item on his plate. Needless to say, this recruit spent the rest of the morning washing about 200 dishes.

After eating, the drill instructors take their platoon to an open area to assess how much the recruits know about fundamental tasks. Such tasks include drill commands, how to stand in formation, and how to march in groups. Although you are still under stern supervision, the drill instructor will teach you how to do specific things necessary to complete training.

The first set of tasks may seem simple, but you might have difficulty learning or remembering them due to the highly stressful environment. If you have trouble learning, do not worry too much because humans normally do not to learn well while under stress. To help alleviate this problem, it is imperative to study your IET book and learn military tasks before basic training begins.

Common tasks will be taught in the field, or in classrooms. You will mostly train for these tasks as an entire platoon, but sometimes you are separated into a squad of about 10 people.

When learning a task that you must perform, the drill sergeant will always demonstrate how the task is done correctly. Pay close attention to these live demonstrations. The activities during Phase 1 are purposely not difficult because the drill instructors know you are pressured. You may not know it, but the sergeants closely monitoring your stress level. The sergeants know when they are providing too much stress, but you will never know this by their actions.

Common tasks you might be taught, during your first phase are as follows:

- Core Values—Respect, Loyalty, Duty, Honor, Integrity, Courage and Service.*
- Drill and ceremony*
- First Aid
- Hygiene/health
- Traditions and customs
- Military law
- Map reading*
- Conduct
- Physical training*
- Guard duty
- Saluting
- Ranks and chain of command*
- General Orders*
- Military alphabet*
- Drug prevention

*Denotes that you should memorize, or be highly familiar with this concept.

Those Boring Films

Drill sergeants know you need time to adjust to the extreme stress of training, but they will make you think that you have no time. This is why for the first few weeks tasks are not too difficult and the sergeants show boring films on obvious topics. Your entire training company (200 recruits) is crammed close together in a warm room to watch films and listen to people drone on about things as exciting as watching paint dry. While you are sitting in metal chairs listening to a topic, your eyes will grow heavy. The only thing you can think of is sleep! However, DO NOT CLOSE YOUR EYES EVEN FOR A MOMENT. Drill sergeants have purposely created a drowsy environment to catch you. Instructors love to knock trainees upside their heads when they find them sleeping. More likely, sergeants will haul recruits who have nodded off into the back of the room. For example, a recruit who was caught with his eyes closed during a video was brought to the back of the room and made to kick himself in the butt with both feet while jumping (called the donkey kick). This punishment exercise sounds funny (it is), but it is very unpleasant after about two minutes.

PT and Your PT test

During Phase 1 recruits will learn how to properly perform military physical training (PT). The military has specific physical training formations and procedures. It is quite remarkable to watch 50 people form perfect lines and perform snappy synchronized exercises. You will do jumping jacks, push-ups, sit-ups, flutter kicks, stretching and will jog in tight formations. Although your 4:30 a.m. physical training takes place nearly every morning, you will alternate exercises which build endurance and strength.

The jogging program separates recruits into ability groups. Naturally slower, or out-of-shape runners separate into a group of runners of similar ability. Usually, the other two groups are the mid-level runners and those runners who should be in the Olympics. A word of caution: do not try to hide in an ability group which is far below your actual ability. If caught, you will have to do so much extra running that your shoes could wear out. If you are a skilled runner, do not have to show off and stay somewhat reserved.

Your PT uniform is basic gray and you are allowed to wear personal running shoes. The military finally wised up and stopped the practice of running recruits in those awful combat boots. Be advised that foot injuries are a leading reason trainees are dropped from basic training. Take care of your feet!

Even though morning PT can be used by the sergeants to punish (motivate) the platoon, it does prepare you for your physical test. Recruits are first given a diagnostic test to measure their baseline physical performance. Usually, this first test is unrealistically hard and it is not uncommon for push-ups and sit-ups not to be counted because they were done incorrectly. My advice on the diagnostic test is make your minimum points in each event, but do not max out. You should roughly do a mid-level range of your maximum ability. For example, if you can actually do 40 sit-ups in two minutes, I would do 35 for the test. By reserving a little of your true ability (assuming you can easily meet the minimum requirements), you buy yourself insurance because you know you can show the drill sergeants improvement when necessary. I am not advocating slacking as long as you know you are in good shape and getting into better shape. However, the first PT is designed to make you think you are in worse shape than is actually the case (Re-read #4 page 89).

Your PT test will consist of three events: push-ups, sit-ups and a two-mile run (three miles for the Marines). The events will always be given in this previous order. The first two events are timed for two minutes. The run is also timed. Make certain you know the minimum requirements for your age before you start basic training. During each phase, you are given a PT test. Only your last PT test is for the official record. The other tests are designed to show recruits progress--- that hard work pays off and to build confidence.

Is It Motivation or Punishment?

Normally, people associate motivation with positive things. Money and positive praise are some commonly used motivators. However, in the military the term "motivation" has a starkly different meaning. To motivate recruits, drill instructors simply use unpleasant reinforcers. Negative reinforcement is when a person's behavior is shaped by activities that they do not want to do again. Trainees will associate negative activities with things they do wrong. Typically, the negative reinforcement exercises are so uncomfortable and memorable a recruit will do whatever it takes to do the assigned task correctly—thus behavior is changed. For example, after doing 50 push-ups and flutter kicks for having a bootlace out of place, a recruit is said to be "motivated" to have the lace tucked properly.

The types of exercises the sergeants use to motivate (punish) recruits range from common push-ups to exotic kicking workouts. There is some debate whether a sergeant can order you to do exercises until muscle failure (the muscle simply cannot do any more movements due to exhaustion), but the sergeants will push you right to that limit. The main purpose of being "dropped" or "smoked" for the slightest mistake is not just so you learn to respect orders, but to get you in physical shape. Even if you did not have morning PT, all the intense motivating exercises provided by the sergeants will certainly get you into top condition. Next time you are being "motivated," focus on the positive fact that you are getting into the best shape of your life.

Why All the Attention On Polished Boots

One task which I initially thought was odd was polishing boots. Maybe polishing boots is not too strange, but when you are ordered to do so twice a day it seems excessive. Why do the sergeants focus so much on the shine of your boots? Obviously, new recruits are being trained to look clean, sharp and focused on details. Many recruits come from backgrounds where it is acceptable to look unkempt, but this is certainly not the case in the military. However, it appeared to me that another major reason drill instructors focused on polished boots is because this task kept recruits very busy. Polishing boots sometimes acts as "busy work" to give sergeants time to relax. Remember, there are 200 trainees in your company and they all have to be actively engaged. There can be no feeling of free time during the first few weeks of basic training.

Other Important Notes Regarding Phase 1

The following are random notes which might be important to remember.

- You will learn different marching songs during all phases of basic. These marching songs are called "cadences" and are designed to build unit spirit. Cadences on long marches or workouts helps to take your mind off the pain produced by the task. These songs also keep every member of your platoon "in step" with each other. I enjoyed the colorful cadences—some were perversely humorous.
- You will receive medical shots and immunizations during your first few days. Do not be alarmed if you feel a little sick, or have a runny nose. It is typical to have minor reactions to the shots. Reactions usually disappear within one to two days.
- Your platoon will have three drill sergeants assigned to it. There is a lead sergeant and two assistants. This three-member team is called a "cadre" (pronounced cod-dray). Sometimes there is a drill sergeant who is portrayed as the nice one, while the other one may be the enforcer type.

- Before you reach Phase 1, recruits should know how to properly address a non-commissioned officer (a sergeant). Never address a drill instructor as "sir," or just sergeant. Always make your replies short and call your drill instructor "drill sergeant."
- Always run everywhere you go. If a drill sergeant sees a recruit walking, that recruit will be in for a long physical workout. You must *look* busy at all times.
- At the end of Phase 1, recruits are introduced to the "low crawl." Low crawling is when a person presses his/her body against the ground and drags him/herself across an area. This technique is very useful in a combat situation, but it is also unpleasant. Dragging yourself over a field produces scrapes and bruises, not to mention a few insect bites. I mention the low crawl because it is sometimes used by the sergeants to motivate trainees.
- From the first day down range, you are assigned to a fellow recruit called a "battle buddy." Personally, I think the term battle buddy is weird, but you must work together to make it through basic. The battle buddy system can be effective, especially during those times when you need a team effort to complete something important. Although I did not care for my assigned "buddy" because he had a habit of sleeping late, you should use this system as much as you can.
- It is vital that your entire platoon works as a team from day one. Your platoon should agree on an unofficial team leader. The team leader should divide up and assign all the tasks as soon as possible. DO NOT FIGHT, OR BICKER OVER PETTY THINGS. The sooner you embrace the team concept, the easier training will be for everyone.

- If you, or your platoon needs to get up five minutes earlier to complete morning tasks, then do so without hesitation. I found waking up a few minutes early allowed me to shower and shave without waiting in time-consuming lines. You will be pressed for time in the morning, so be efficient.
- As far as privacy is involved, you do not have any. It seems like you are never more than a few feet away from someone else at all times. Get used to it.
- Learn to love water. Water will help you fight fatigue and drowsiness. Water is also very good for you when you are performing intense physical activities. During my training, we drank nothing else but water (no soda, milk, juice). We were ordered to drink a canteen per hour on hot days.
- Learn to pay attention to detail. If you are assigned to do a task, do the task perfectly by attending to the details. Drill sergeants look carefully at the small details to see if you are "squared away."
- Always remember the big three commandments during Phase 1. Never show emotions, do not stand out and do not volunteer for anything.

Phase II (White)

The transition from Phase 1 to Phase 2 is not always readily noticeable. However, some units have a ceremony marked by changing the guidon flag from red to white. A few days into this phase, you will notice that the sergeants are actually instructing more than in Phase 1. Phase 2 is where you will learn meaningful skills and harder tasks. Your PT should become more challenging as well as your drill and ceremony (marching maneuvers). By far, most of your time will be dominated with weapons training, or Basic Rifle Marksmanship (BRM).

To me, Phase 2 seemed the longest, yet most entertaining. Personally, I thought the tasks were more exciting, but also more tiring. For example, we had to march further to our area of operations (AO) in the heat with full combat gear which was fatiguing. However, once at the AO we did novel activities like throwing hand grenades.

Common activities for Phase 2:

Weapons training
Grenades
Camouflage/concealment
Bivouac (camping)
Hand to hand combat
Pugil stick competition
Obstacle course
Hand grenades
Bayonet assault course

How Recruits are Taught

There are several ways to teach new tasks, or to impart knowledge to people. The military utilizes a particular method of teaching which is technically described as "reductionistic." In different terms, a complex task is reduced down into many simpler sub-tasks. Successful completion of the simple sub-tasks will ultimately produce successful completion of the larger tasks.

To illustrate, recruits are taught to disassemble their weapons quickly. To take apart a M-16 rifle, a trainee will have to go through a sequence of steps which must be done in order. As with any new task being taught, the drill sergeant will first demonstrate how the steps are correctly performed. After carefully watching how the steps are performed, the drill sergeant will make recruits practice only one step in the multi-step process. You might practice this one simple step until it becomes painfully boring. However, when a simple task becomes boring, this means you have committed the task to deep memory—which is critical. When an activity is placed in deep memory by repetition, you will likely not forget how to perform the activity even under combat stress. Therefore, learn all tasks to the point that it becomes boring because it is necessary and desirable.

Weapons Training

Once introduced to your weapon, you will spend a very long time with it. You will be issued a M-16A2 assault weapon. Most people find handling a high-powered weapon exciting, but I can not emphasize how fatally vital it is that you are safe with it. I use the term "fatal" because the sole purpose of the M-16 is to efficiently take a human life. While people can survive mistakes with some civilian firearms, a mistake with the M-16 kills on the first error. What kills is when careless trainees point their weapon in the wrong direction, forget to set the safety switch, or leave a bullet in the chamber.

If you listen to the training instructor with the utmost focus, you will learn properly how to handle the M-16. Your first and constant lesson with the M-16 will be safety—learn it well. If you are careless with your M-16, you will regret ever enlisting. I am deadly serious.

My first advice regarding your weapon is to memorize both the serial number and its weapons number. After you memorize the numbers, never let your weapon out of your sight unless ordered to do so! Do not ask others to hold your weapon while you do another task. If you misplace your weapon, you will be held responsible for its replacement. Also, you could do jail time in addition to being dropped from the service. My intent is not to scare you, but rather to make you understand the seriousness of having a weapon issued to you.

Once your safety procedures are ingrained, you will learn how to clean and shoot your weapon. While being safe requires certain procedures, you do not have to handle your M-16 like it is made of glass. Actually, the M-16 is an extraordinarily tough and rugged piece of metal. Do not be afraid of your rifle, especially when shooting it. The rifle is lightweight, easy to aim due to large sights, and highly accurate. It is not hard to become a marksman with this rifle because it is so easy to aim and fire. The best aspect of the M-16 is that it does not "kick" much at all.

The worse aspect of the M-16 is cleaning it after firing. Recruits spend hours cleaning their weapons. You will be given the required cleaning tools and shown exactly how to use them. Also, you will be shown how to take apart the M-16. You will be amazed how many tiny parts are built into this rifle.

Insider tip: The most important tips to keep in mind when cleaning your weapon are to put the smallest parts (springs and pins) on a clean small towel and keep those parts visible at all times. Most recruits have difficulty taking off the two black plastic hand (barrel) guards. You have to pull down on a holding ring at the base of the barrel to remove the guards, but on some rifles this is difficult. I used a long metal cleaning tool to take my guards off and put them on properly.

The last note I have regarding your weapon is how you refer to it. The M-16 is a weapon or rifle, NOT a gun. Do not refer to your weapon as a *gun* or you might find yourself singing a famous vulgar song (from a movie called "Full Metal Jacket") and doing push-ups.

The Firing Ranges

I think most recruits enjoy the thrill of firing their high-powered weapon. If you relax with it while shooting (but not with safety), you will find that the firing ranges can be fun. You will shoot hundreds of rounds of ammunition. In fact, recruits shoot so many bullets that even the worse shooter cannot help but to become a good marksman. The reason the military spends the majority of time on weapons is because all personnel must know how to fight. After all, the military would not be a fighting force if soldiers do not know how to shoot. Everyone at the basic level is a warrior.

A special instructor called the range master strictly controls the firing ranges. The range master announces over a loud speaker precisely what you should be doing at any given time. He/she will tell recruits when to load their weapon, when to aim them, when to move the safety switch, when to fire and when to stop. Listen to the range master like your life depends on it.

Most ranges I went to in basic were filled with high technology. The ranges are quite large and the area where the targets are located extends farther than you can see. Usually the last target has an earthen berm to stop the bullets. There are several "foxholes" in a straight line for shooters to stand in while aiming their rifles down range. Some ranges have moving targets, while others pop up. My range was electronically controlled so that when a recruit hit a target, the target fell and a computer kept score.

Another important thing to remember under this section is do not attempt to take any ammunition from your area (called brass, or ammo). You will be checked as you leave the firing range. Also, do not carelessly leave a bullet in the magazine or chamber as you leave the range. A recruit in our platoon accidentally left a bullet in his rifle and it went off while he was returning the weapon. This very serious incident punished the whole platoon for a week. The entire platoon was ordered to perform perimeter guard duty which meant that few of us received more than a few hours of sleep for a long time.

Qualifying with Your Weapon

All recruits must qualify with their weapon. Qualifying means a recruit must hit a specific number of targets given a certain amount of bullets. For example, you will be given 40 bullets to hit a minimum of 28 targets. 28/40 qualifies you as a basic marksman. If you get more targets, you could make "sharpshooter," or "expert."

Insider's tip. A tip when qualifying is to keep shooting all your ammunition. Sometimes a bullet becomes stuck in the chamber and you have to eject it. Therefore, after you fire all of the bullets in your magazine, reload the ejected bullet and fire it. If you cannot find the ejected bullets, there may be other good bullets near you that you can quickly load.

When aiming at your target, you are supposed to sight the middle (largest) section. However, basic training weapons have been under constant abuse by previous training cycles. Therefore, expect your weapon not to fire as accurately as a new weapon. For example, my weapon shot low, despite having "zeroed" it in several times. I had to compensate for the inaccuracy by aiming at the head of the targets. In other words, it is important that you understand the particular way your weapon fires. Find out if your rifle shoots high, low or straight and adjust accordingly.

Hand-to-Hand Combat

Learning real fighting moves and techniques takes a long time to master. It is doubtful that you will be a "killing machine" after the few hand-to-hand classes taught during basic training. However, if you have a combat MOS, you will learn the deadly arts in advance individual training (AIT). Basic training recruits spend several hours a day for a few days flipping each other and pretending to stomp on each other. The entire company performs combat drills while in formation.

Once hand-to-hand combat is over, your platoon may have pugil stick competitions. Pugil sticks look like huge Q-tips with pads on either end. Recruits are dressed in protective gear and a helmet. Two trainees start out back to back until the drill sergeant blows a whistle. Immediately upon hearing the whistle, the two challengers beat each other senseless with the sticks. Of course no one gets hurt, but points are kept for each recruit's team.

Insider's tip. Most people are not used to this sort of competition. Pugil stick competition is fun, but it is also good practice for using a bayoneted rifle as a weapon. I noticed competitors who ducked frequently and jabbed (thrusting forward, not swinging) won most of their matches.

How Many Platoon Leaders?

You might have a platoon leader assigned during Phase 1, but usually Phase 2 is where platoon leaders are appointed. Platoon leaders are recruits who have adjusted to basic training and who are selected by the drill sergeants to act as a platoon assistant. The drill sergeants will convey messages to the platoon through the leader and assign him/her extra leadership duties. Platoon leaders usually stand in front of the platoon, sometimes giving the platoon orders to march or to do other tasks.

Some recruits look forward to being a platoon leader, while others shy away from it. I have mixed feeling regarding this position. On one hand, it is an honor to be a leader and you do develop some leadership skills. On the other hand, this is basic training and you are given much more responsibility with little or no extra benefits for your hard work. Additionally, you are held accountable for everyone in your platoon. Therefore, if one recruit is late, or messes up on a task, you get to do push-ups until your veins pop out. I also realized late in Phase 2 of my training that the sergeants "fired" platoon leaders regardless of whether they were good, or not. The sergeants wanted to give other people a chance to develop their skills, so we had many fired platoon leaders—including myself. I personally do not think it is wise to volunteer to be a platoon leader until Phase 3, when most recruits are "squared away."

The Feared Gas Chamber

If you do not know anything about the gas chamber, keep it that way until you go through it. If you have heard stories about how painful it is, try to keep your fears in check. The bottom line with the gas chamber activity is that it is very unpleasant, but it is not as bad as you think.

The military has new recruits experience the gas chamber for many reasons. First, chemical/biological warfare is a very real combat threat and is just as serious as nuclear weapons. The most effective way to make soldiers remember survival tasks associated with chemical weapons is to make them experience the harsh effects of such weapons. Second, the military needs recruits to trust their gear designed to protect them. Soldiers will not be focused on their tasks if they think their protective gear cannot protect them adequately. Last, the drill sergeant wants to see which recruits will maintain their military bearing under the intense stress of a combat climate.

Here is how the drill will proceed: Your entire company will arrive at the gas chamber area. A special instructor will explain how to properly wear your protective gear and give a speech about the grotesque effects of real chemical agents. The special instructor will also tell you about what to expect once inside the chamber.

While some platoons are waiting to enter the small cinderblock gas house, some platoons will be training for nuclear warfare nearby. You will learn how to protect yourself from radiation and blast effects. However, most of the nuclear training entails diving to the ground and rolling around in sand. The entire

NBC (nuclear, biological, chemical) training is largely obnoxious, but it ends in a few hours.

After your nuclear training, your platoon stands in line before the gas chamber which resembles a large brick tool shed with small windows. As you wait, you will notice the door on the other side of the house fly open as recruits from inside pour out gasping for air. Recruits just experiencing the chamber will have fluids draining from every opening on their face. Eyes are red and mouths have long streams of saliva running down. I think the drill sergeants purposely want the waiting recruits to watch the pain of the other trainees. The sergeants know they can make recruits more fearful if they can get the recruits' minds focused on the suffering trainees leaving the chamber.

As you are pushed into the gas chamber by the drill sergeant who guards the entrance, you will be wearing your gas mask. If your mask is put on correctly, you will not feel the gas at all. (Ask your neighboring recruit to double check your mask before you enter the chamber.) Once inside, there will be a small centered table with a canister on top pumping out CS gas. The drill sergeants inside the chamber are wearing their masks and they make the recruits surround the small table. When the door shuts tight, the room becomes much darker and it quickly fills up with a gas fog. At this point, you do not feel anything because you are protected. Next, the sergeants inside the room order you to repeat your roster number as loud as you can. Then, the instructors order you to remove your mask. If you do not remove your mask, a very angry drill sergeant will yank it off of you.

Once your mask is off, instantaneously the CS gas feels like very hot oven air against your face. The first thing the gas stops is your breathing. When your breathing stops, this will scare you like nothing you have experienced. Your eyes automatically shut tight despite your hardest efforts to open them. Tears and mucous will immediately flow from your face like a faucet. Suddenly, the back door opens and all the recruits run out in an orderly fashion until they are about 10 feet away from the building.

As you run out of the building gagging and in obvious pain, you think the drill sergeants will give you a few seconds to collect yourself. Actually, the drill instructors are very stern after you leave the chamber. I remember our drill sergeant gave the ridiculous order to stop coughing, which was an absolute biological impossibility. Because some recruits were still wheezing, the drill sergeant brought us to the sand pit for a sweaty roll in the muck.

Insider tips for the gas chamber. Your mental images and fear will be your worse enemy for this activity. As just described, it will be painful, but not deadly. The effects of the gas instantly start to decrease once you are back outside. You *must* tell yourself, "This is going to be painful, but I can do it. It is not going to kill me." You cannot get out of this task, so power through it with your positive attitude.

Once inside the chamber, before you take off your mask, remain *calm* and *relaxed*. Focus your inner speech on how well your mask works and how goofy everyone looks with their gear. Find humor in the chamber and keep telling yourself, "Hey, this isn't bad. Besides, I need my nose cleaned out." Once the order to remove your mask is given, take a deep breath. Close your eyes *first* and then remove your mask. You must look as normal as possible, do not look like you are holding your breath. You will quickly feel the gas stinging you. As you very slowly release the big gulp of air you are holding, start to take *shallow* breaths. If you have to repeat your roster number, move your mouth, but just whisper the number. Do not listen, or watch other trainees. Keep telling yourself the gas is not too bad and keep your eyes shut. Although it will feel like a long time, the exit doors will open and you will be outside in about a minute. Remember, once outside do

NOT expect any sympathy from the drill sergeants. Maintain your military bearing.

Throwing Grenades

Every trainee seems to like this brief exercise. Like most training tasks, you will arrive at a specific training area and wait for the special instructor. The special sergeant will demonstrate exactly what to do and what not to do. Anytime these instructional sergeants speak, pay extra attention and focus because they rarely repeat themselves.

In an adjacent area, you will practice throwing dummy grenades. Once you are "good to go" as the sergeants say, your platoon will stand behind a thick wall. As you wait to throw your two real grenades, you can hear the detonations of those grenades thrown by other trainees. The sound of exploding grenades is a very loud, but deep sound. The vibrations go right through the thick protective wall. Your line of fellow trainees will slowly draw closer to a sergeant who stands at the entrance to the grenade field. The entrance sergeant gives you a heavy flak jacket to wear and looks you square in the eyes. Once he feels you are not scared, he hands you two baseball sized metal spheres. You run with the two grenades close to your chest to an awaiting drill sergeant who is standing in a deep pit. This safety sergeant is extremely serious because mistakes here can cost him his life and yours. You are ordered to pull the grenade's safety pin and then ordered to throw. Immediately after you throw the grenade, you dive to the ground.

Insider tip. Once you are ordered to throw your grenade, throw it as far as you can. When the grenade is released from your hand, instantaneously hit the ground. DO NOT TRY TO WATCH WHERE THE GRENADE GOES AND DO NOT TRY TO WATCH THE EXPLOSION. If you do not hit the bottom of your safety pit immediately after you throw your grenade, the 250-

pound sergeant will slam you into the ground. Then, the sergeant will make you regret your next breath. In other words, do not screw up during grenade throwing because it is just as deadly serious as shooting your M-16.

Map Reading

Military map reading is not like reading a street map. Usually, basic map reading entails understanding topographical (topo) maps. A topo map is a special map which illustrates mountains, canyons, rivers, lakes, forests and roads in a particular way. Topo maps are filled with small lines which show the height of the represented area.

Map reading in basic training is elementary and not too difficult if you have been exposed to these maps beforehand. Most recruits have trouble with the new geographic words such as saddle, spur and contour interval. For your individual task test, you will be shown a topo map and asked to point to specific features which have been taught by a drill instructor. Also, you will have to understand compass directions as well as measure distances on a map. Again, these land navigational tasks are taught by special instructors, so pay close attention.

Insider tip. If you have time before shipping out for training, visit your local sporting goods store. Ask the clerk to see a topographical map of a mountain state like Colorado. Mountain state maps usually have all of the features you are required to learn. Topo maps are not generally expensive. If you can afford one of these maps, buy it and study it carefully. If you do not buy a topo map before basic, study your IET book and memorize the geographical terms it provides. You should know what the following map features are: saddle, spur, depression, hill, cliff, draw, ridge and valley.

D and C

D and C is known as "drill and ceremony". D and C encompasses fancy marching, formations, and inspection procedures. From the time you arrive at basic, you will spend hours learning how to march in precise unison as a large group. After you have mastered synchronized marching, more sophisticated maneuvers are taught. For example, 40 recruits are ordered to "Left . . . Face." Instantly, all the recruits act as one and face to the left. Then, another command is barked out "Forward . . . March." Obviously, the whole platoon takes its first step exactly at the same time and moves forward. Soon, other orders are yelled out in rapid succession such as, "Half step. . . March," then "Rear face. . . March."

The primary reason behind marching is to efficiently and orderly move huge numbers of soldiers from one area to another. However, a secondary benefit of D and C is that unified marching is an impressive ceremonial activity which builds a sense of order, discipline and morale. Indeed, D and C is an amazing sight when executed in an abrupt military fashion.

Insider tip. The biggest hint about D and C is to listen to the *preparatory* command before the actual execution of the command. For example, when the sergeant yells, "Forward...March" the three dots after the preparatory word "Forward" indicate a slight pause before the *execution* word which is "March." Do not change your current movement, or stance until the execution word is given. Once "March" is ordered, perform the command instantly *after* it is said. Only act after the pause and on the *second* word command.

You must perform all D and C procedures in a snappy, quick manner. If you can, practice marching and other D and C maneuvers before you ship out. Ask your recruiter to help you learn some of the commands. Also, you will have a brief few days to practice before training at reception battalion.

Another hint is never move when the order "Attention" is given. When recruits are standing at the position of attention, they must stand rigidly straight with their hands to their sides. Eyes and head must be fixed straight and level. You must be a concrete statue and not move any muscle while in the attention position. Even if a wasp is taking chunks of flesh from your neck, do not move. The most common excuse drill sergeants have for dishing out push-ups is because some trainee wiped sweat from his/her eyes during the position of attention.

Memorize the following commands and fully understand how to execute them. (Note, the dash (-) within words denotes a quick pause in how the commands are given, or the preparatory and execution part of the command.)

Common D and C Commands:

About-face
Stand-at-ease
Parade-Rest
Double-time-march
Present-arms
At-close-intervals
Atten-tion
Counter Column-March
Halt
Quick time-march
Dress-right-dress
Inspection-arms
At-ease
Rest
Port-arms

(See next page for other commands.)

Inspections

Recruits will have many inspections throughout each phase of training. It seems Phase 2 is where inspections become more genuine and not just a reason to administer "motivation" exercises as seen in Phase 1. Usually, recruits stand in formation at the position of attention and await a drill sergeant or company commander to inspect them. High-level sergeants or officers examine how you are dressed, your neatness, how polished your boots are, the cleanliness of your gear and how you act under the scrutiny of being inspected. Many times, the inspecting sergeant will ask you questions about the chain of command, details about your weapon and details about military procedures. You will have learned everything the inspecting officer asks, but you will be nervous and might forget. If you remain relaxed and focused, you will most likely answer questions correctly.

"Inspection Arms" is a special type of inspection involving your M-16. Your IET handbook outlines six steps of Inspection Arms which you will practice for hours during training. The primary tip to remember during this special inspection is to perform all movements in an extremely quick, but correct manner. For example, when the preparatory order is given, "Inspection...," get ready to grab and bring your weapon to its up position in a crisp manner when the execution command "Arms!" is given (the full command is, "Inspection...Arms").

Most recruits have difficulty with presenting their weapon to the inspecting officer without looking down. Recruits must learn inspection arms without looking. The hardest part of inspection arms is pulling the charging handle of the M-16 back to expose the rifle's empty chamber. The only way to become proficient at this task is to practice it until it becomes automatic. Additionally, always present a highly polished, clean and *unloaded* weapon to the inspecting person.

Do not be surprised if you fail all of your first inspection: that is part of the game to build confidence. However, fair inspections will be done late in Phase 2 and in Phase 3.

Radio Communication

Recruits learn how to use combat communication devices which require specific military procedures. Recruits must memorize and properly use the military alphabet and radio code words.

Once you know how to "radio speak", you will be ordered to make a report for a test.

Most recruits will learn to use the SALUTE method to report on enemy observations. The letters of SALUTE stand for: Size, Activity, Location, Unit, Time, and Equipment of the enemy. Trainees learn the proper military protocols and terms when using tactical field radios. A drill sergeant will be on another radio asking you quick question regarding SALUTE. I noticed some recruits had great difficulty with this task because they did not understand the military alphabet, or how to start communications. Again, if you memorize the radio procedures given in the IET book beforehand, you will perform this task sufficiently.

My Feet are Aching-- Road Marches

Interestingly, I could run five miles with little problem. Running was usually easy for me. However, give me the same five miles as a road march and I struggled. You will find some people are good at both running and road marching, but most of the time people had their strength in only one of these tasks.

Road marching is where recruits dress in full combat gear, rucksack, weapon, ALICE belt and canteen full of water. Trainees march with their gear in long lines down the sides of roads. It may not sound difficult, but the gear you wear is heavy, your boots are stiff and your uniform makes you extraordinarily hot in the summer. Furthermore, recruits have to keep in pace with each other, which is not easy to do because some people have longer strides than others. If you do not keep the proper pace and spacing between each other, the sergeants will pull you out and make you run around the entire formation with full gear. Sometimes for punishment, recruits have to run while holding their M-16 high above their head with both hands.

During Phase 1, road marches are not really tough at all. Usually, sergeants might march you across a field with no gear.

However, during Phase 2 recruits start to learn real road marching. Initially, road marches are very short, but quickly move up to a few miles.

Insider tip. At all times during training, take good care of your feet. Trim your toenails constantly and keep your feet dry with powder. Massage your feet at night, or soak them in very cold water. You should buy high quality thin insoles to put into your boots before road marches. If you know a road march is the next day, drink as much water as you can the night before you start. Drinking large amounts of water is called "hydration" and is extremely important to your performance and survival. It is common to drink about a quart of water for every 1.5 miles you march in the summer.

Another good tip is to involve yourself in the marching songs, or cadences. These songs can really help take your mind off of your swelling feet. Also, I used mental imagery to help take my mind off of the discomfort. By vividly picturing myself on a beach drinking an ice-cold beer and listening to my favorite songs, the time passed quicker than if I had just focused on how much longer we had to march. Use your imagination, or whatever else works to defocus the pain.

Eating Chow and KP

Recruits are well-fed during basic training. Three times a day, trainees will be fed a balanced meal either in the field, or in the cafeteria (which is also called the defac, mess hall, or chow hall). The food is not like it was 30 years ago when it had a reputation for tasting like cardboard. In my opinion, the food is somewhat tasty, but you do not get to enjoy it much during the early stages of training. The amount of calories people burn during training is very high, so most recruits at this point try to eat as often as possible. If you have a weight problem before training, by the third week your problem will not exist due to the long physical days.

In Phase 1, recruits form a tight line before the cafeteria's entrance. There is absolutely no talking allowed and recruits must stand at attention. When recruits enter the dining facility, they hold up their identification card and yell out their number. After you sign a roster, you move like a robot through the food service line. You must eat everything you are served. Recruits are carefully supervised and forced to eat within just a few short minutes.

In Phase 2, eating becomes less stressful and will remain the same until you finish training. Usually, recruits are marched back to the cafeteria after completing field activities. The long line before the entrance may not be as strictly supervised, but you might have to do 10 chin-ups before entering the facility. Also, during these later phases, drill sergeants may quiz you on parts of your IET book before you can eat—which is why I suggest you memorize key parts of it early.

Field chow is served in Phases 2 and 3 and it is just as good as the cafeteria's food. Typically, active duty cooks bring special trucks and containers full of hot food to the field. If you are eating in the field, such as at a firing range, make sure you sit well away from the drill instructors. Drill sergeants have a habit of selecting nearby recruits for kitchen duty.

KP, or kitchen patrol, is when a trainee is assigned to work in food service. Every recruit will be assigned KP at least once during training. Some recruits enjoy working in the kitchen because they eat early and receive extra rations. Also, recruits get to escape the constant yelling of the sergeants for the day. However, I found KP very hard and hot work. On KP, you will wake up earlier and work longer than your typical day. You might find yourself scrubbing floors, loading tons of nasty garbage, or washing a mountain of dishes.

Phase 3 (Blue Phase)

You are almost finished training and you can sense the end is near during this phase. The drill sergeants give recruits more personal time and tasks are taught in a straightforward manner. Most activities are no longer designed to punish recruits, but are intended to build confidence. You might see the drill sergeants assisting and speaking to recruits more than in the other phases. The drill sergeants' "Smoky the Bear" hats are replaced by the same headgear recruits wear. It is hoped that trainees begin to see that they are becoming more like their sergeants not only in appearance, but also in attitude. The main idea during Phase 3 is that recruits are beginning to earn respect and are no longer seen as cattle by the drill instructors.

Still, there are some major tasks associated with this phase which must be successfully completed. Recruits must complete a final road march, a long run, a field training exercise (FTX) and pass their final official PT test. For other branches such as the Marines, there is a water survival task and an intense three-day exercise called the Crucible.

The overall advice I have for recruits for this last phase is give your best effort. Make sure you work as a team for every task and keep in-fighting to a minimum. Most important, do not become careless and lose your military bearing. Some recruits during this phase try to buddy up to the drill sergeants because the sergeants *seem* nicer now. I suggest you remain focused and serious. Do not be surprised, even at this phase, if a drill instructor reverses back to a Phase 1 attitude.

Field Training Exercise (FTX)

There is minor difference between a bivouac and a field training exercise. Although both exercises involve camping in the field, the FTX is much more combat- oriented. The Phase 3 FTX involves tactical maneuvers, live-fire exercises and combat simulations. FTX can last from a few days to a week In contrast, during the Phase 2 bivouac, recruits learn to set up tents and perform field duties in a non-combat-like environment.

During a FTX, your company will set up tents and dig latrines. Next, you prepare for night firing exercises which involve flares and live ammunition. You may even use night-vision goggles and fire your weapon on full automatic mode. Trainees learn combat soldiering skills like the three-second rush, camouflage and how to deploy mines.

In the field, recruits are usually fed MREs (meals ready to eat). These packaged meals come in a single heavy-duty plastic bag and are complete with everything you need to eat a hot meal. Not all MREs are alike (as denoted by a number on the front of the brown bag). Some are much better than others, but all have the necessary calories to sustain an active soldier for an entire day. Most soldiers even ate MRE's that they did not like because basic training makes you extremely hungry. You will be amazed at what you will eat when you are burning over 2000 calories per day and feel starved. No one has a weight problem by Phase 3 and you will eat everything inside your MRE pouch. I even saw hungry recruits eat the coffee grinds out of their packet.

In the end, FTX is fun and you begin to feel like a real soldier. Pay attention to the skills you learn at your FTX. Enjoy yourself, but keep a serious attitude.

Live Fire Exercise

Sometimes during the FTX, a live fire exercise is planned at night. During my night-fire course, we had to crawl under barbed wire while real bullets whizzed over our heads. The bullets were coated with phosphorus which made them streak by with intense light. These tracer bullets lit up the dark sky like speeding fireflies. The unusual glow of the tracers created an optical illusion which was very strange to see. Bullets that seemed only inches away from my face were actually several feet above me. Regardless, it is important to strictly adhere to all safety procedures during any live fire exercise and keep your head down.

Obstacle and Confidence Courses

Depending on your training company, you might complete a special obstacle course called the confidence course during the early part of Phase 3. The confidence course is a series of challenging obstacles you must climb over to finish. Typically, you are put into a team and your team helps each other navigate the course's structures. Collaborative effort is a key point the sergeants want you to learn on the confidence course. Some obstacles can be entertaining, but you will need your team's assistance. Recruits will have to climb towers, swing across water, repel down ropes, or slide down cables using a pulley-like device.

Another obstacle course your platoon might encounter is the conditioning course. This course is a timed event and focuses on the individual trainee. Trainees have to scale low walls, climb ropes, crawl through tunnels and traverse monkey bars. If you do not complete the course quick enough for the sergeants, they will make you do the course again. The first time is tiring; the second and third time are exhausting.

All of these courses are rewarding to complete. For people afraid of heights, or not accustomed to climbing, completion of these courses certainly builds pride.

Bayonet Assault Course

Again, depending on your training company, you may complete the bayonet course in the second or third phase. I had a feeling that our training cycle was behind schedule, so we completed the bayonet assault course during the first part of Phase 3.

Recruits are given a 10-inch steel blade to affix to their M-16 rifle. The M-16's bayonet is an extremely rugged and durable piece of metal which looks exceptionally lethal. Essentially, this incredibly sharp blade turns the M-16 into a spear, but you do not throw it.

Usually, the course is set in a muddy field with obstacles and life-sized dummies. Some dummies look like mannequins, while others are just old car tires attached to a wood post. Wearing leather gloves, recruits are sent one by one charging toward a target. You are supposed to yell like a crazed escapee from a mental hospital while you forcefully stab your target. Drill sergeants will be screaming at you to be louder and more violent. I remember one part of our course was flooded with warm muddy water. As recruits approached this area, the drill instructor told them to low crawl through the mucky soup. Recruits who were not totally soaked in mud were sent back for a re-dunking.

For some reason, this particular activity is where some recruits receive the most scratches and bruises. Even though recruits wear heavy leather gloves, they can still manage to hurt themselves at the course, so be cautious.

Your Final PT Test

Unlike your first PT test during Phase 1, your drill instructors are sincerely supportive at this point. The motivation the sergeants use in Phase 3 is more positive than negative. Still, do not expect any praise and you won't get upset. Furthermore, the drill sergeants actually count most of your sit-ups and push-ups, unlike Phase 1. The contrast between Phase 1 and the last phase is to exaggerate the progress you have made over the last few months. However, due to the intense exercises and PT you have been doing since week one, all recruits will show a significant improvement in their PT scores. Without a doubt, all recruits are stronger and faster than when they first arrived.

For their final record, recruits are given the big three timed events: push-ups, sit-ups, and a two-mile run. Do the absolute best you can for this record.

End of Phase Testing

All three phases of training will have performance tests. The first series of tests in Phase 1 are simple, such as how to salute, or how to put on your gas mask within a few seconds. In other phases, the tasks are a little more difficult to perform and the standards to pass are a little stricter. Usually, one of your three drill sergeants will test you on a specific military activity. Sometimes you have to perform the task alone, but other times you complete tasks in a small group. If you pass your required assignment, the drill sergeant will mark on your record "go." If you fail, your record will have a "no-go." If you get a "no-go," you will have to practice this task until you are proficient. During the first few phases, if you get a "no-go" you are generally motivated with punishment exercises. During the last phase, the instructors actually attempt to remediate your skills by teaching you the skill. Your end of cycle test (EOCT) is for your official record, so give your best effort.

The Last Road March

Of all the events, I dreaded the final road march. I would have gone through the gas chamber again instead of this hike. Not that road marching is hard for everyone; I just personally disliked this activity. Because my natural gait is short and the heat was intense, I knew the march would last several unpleasant hours.

Most final road marches are about 10-12 miles. Over the last few weeks, recruits are slowly conditioned to long road marches. Normally, the sergeant will not exercise the platoon too hard the day before and will skip morning PT the day of the march. Starting before 4:00 in the morning, your training company hits the road with full gear. Your drill sergeant constantly makes sure you keep the proper pace and you must carry your weapon at the proper “ready” position. The weapon, which weighs about 7 pounds, begins to feel like 20 pounds after the first few miles. At regular intervals, your company is allowed a few minutes to rest and drink water. You are constantly told to down as much water as you can.

If you keep your mind busy and sing the cadences, you will finish. Even though your feet will be swollen and hurting, the feeling you have when you complete this march is like winning the lottery.

Gear Check and Graduation

After your final inspection during Phase 3, your sergeant will examine all of your gear. The gear you were issued during reception battalion is yours to keep. However, all the equipment you received down range at the training barracks stays there. Recruits do not keep helmets, rucksacks, canteens or any type of weapon.

The last week of training is actually quite busy. Recruits now spend their day cleaning gear and polishing the barracks. Moreover, you are issued your “Class A” dress uniform and take an official picture while wearing it. Your "Class A" uniform is a green polyester uniform used for special military functions. Even though I do not like the look of it, dressing in my Class A uniform at the end of basic brought a deep sense of pride.

The last day of basic training produces the most positive emotion humanly possible. Not only are you tremendously elated that boot camp is finally over, you will feel an incredible sense of achievement. You will start your graduation day with drill and ceremony and a speech by the company commander. Then, as your family and friends watch, you are dismissed from basic. Your family and friends can join you at the base’s dining area.

Now, you can speak freely with your drill sergeants, or even take a picture with them. It is finally over!

A special note: Reservists and National Guard members who split their training between two summers (basic training one summer and advanced training another) typically do not have a graduation ceremony like the one described above.

On to Advance Individual Training (AIT)

Some recruits will complete both their basic training and advance training at one site. Many times, recruits at these One Station Unit Training (OSUT) sites will have the same drill instructors—especially if they are in a combat arms MOS. Therefore, when basic is completed at OSUT sites, only a small ceremony is performed and then advance training begins. Typically, the company commander gives a few days break to recruits before AIT starts.

In contrast to OSUT, some units have to send their basic training graduates to a different military base. After a brief few days of break, it is common that new graduates are flown to different states for AIT. When new trainees begin their advanced training, the focus is on developing meaningful skills they need to learn for their MOS. Life in AIT is a little more relaxed compared to basic training. However, sometimes the instructors like to remind trainees they have not completely finished all of their training and will "motivate" the platoon.

Chapter Summary

1. The following is the general structure of training:

1. Arrive at reception battalion.
2. Start Phase 1 (three weeks long)
3. Phase 2 (three weeks)
4. Phase 3 (three weeks)

5. Graduation from basic training
6. Continue with, or ship to AIT

2. Reception battalion is where thousands of recruits arrive and are "in-processed." Records and medical examinations are given. New gear is issued to all trainees. Activities at reception battalion focus on preparing the recruit to start basic training which is at a different location on the base. While strict, the stress of reception battalion decreases in just a few days. You should focus on learning your IET book and drill procedures.

3. You are shipped from reception battalion to your basic training site in hot, cramped cattle cars. For the next several hours, you will experience extreme stress and confusion as the drill sergeants attempt to establish total control.

4. Drill sergeants use mind games to control and train recruits. You should be aware of the psychology behind many techniques and keep a positive attitude about training in spite of the stress.

5. Phase 1 is the most difficult period where recruits learn to manage stress. Do not expect to do anything right, so just go with the flow of things. Remember the three major suggestions: do not stand out; do not volunteer; and do not show emotions.

6. Phase 2 is where you start learning more meaningful tasks. Some tasks like weapons training can be fun. Try to view as many tasks as possible as fun, or as a good learning experience. During this phase, drill sergeants have a habit of being nice one day, and harsh the next. Always remain serious, but relax when you can.

7. Look for hidden downtime to relax and de-stress. Sometimes when cleaning your weapon, you can find time to relax. Speak with friends, use self-talk and see the humor in things to fight stress. Do whatever it takes to remain positive.

8. Always be extra careful with your weapon. Maintain your weapon properly, watch its parts closely and be safe at all times. Never let your weapon out of your sight and memorize its numbers.
9. Phase 3 is where you should give your best effort. You can relax more in this phase because the drill sergeants want you to start feeling good about yourself. You will have more personal time during this phase. You will be excited that graduation is near, but do not lose your military bearing and focus.
10. Despite your attitude about basic, or the military, you will feel a strong sense of accomplishment and pride once you complete training. After a ceremony marking the completion of basic training, recruits continue with their advanced individual training called AIT.
11. Some recruits will stay on their basic training base for AIT, while some leave for other bases. AIT is more like a school and less strict than basic training. Still it is a good idea to maintain a focused attitude during AIT.

Chapter 8

Additional Aspects to Consider

I purposely separated the following points from the previous chapter because they deserve special consideration. Some of the following items (like an Article 15) most people will not experience. However, it is important to understand the following because you will hear about them during training.

Sick Call

Every soldier has a right to receive medical attention and to go on sick call. Many times, drill sergeants will discourage recruits from going on sick call. In fact, in my experience, the sergeants create a mildly harassing climate for those who attempt to see the doctor. For example, recruits who were put on the injured list were called "broke dicks" in my company. In spite of the discouragement you might receive, I strongly suggest that you see a medical specialist if you are hurt.

The following procedure demonstrates how to properly seek medical assistance: You report to the platoon leader or drill sergeant that you have an injury. You then get permission to go on sick call. Usually, the next morning at first formation (4:30 a.m.) you will stand in a small "sick call" group apart from the company's formation. A vehicle usually brings the sick call group to the infirmary, but other times the group walks to the medical facility. Recruits complete a short medical form and then wait to see the doctor, physician's assistant, or medic. Medical services are free to all military personnel.

Although I advise recruits to see a doctor when they are injured, some trainees fake an injury. A few misguided people attempt to skip training by hiding within the sick call ranks. In time, these fakers are discovered. You do not want to be a faker because such people become the favorite verbal punching bag of the drill sergeants.

Religious Services

The military makes an impressive attempt to honor all religions. All recruits have the right to practice and observe their religious beliefs. A military chaplain is a highly trained military professional who is well-versed in many major religions. Typically, recruits will observe their beliefs on Sunday. Recruits are allowed to attend services at a multi-denominational facility (sometimes referred to as "church"). I advise recruits who are not religious to attend services on Sunday because it is a "free hour." Especially in the early phases of training, you can actually relax during this religious hour. Besides, if you decide not to attend services, you could find yourself working as a cleaning crew member around the base. As Phase 3 nears and the sergeants are a little less strict, you might find other ways to spend your personal time besides attending services.

Chaplains and Counselors

As mentioned in the previous paragraph, chaplains are highly trained religious professionals. Chaplains are interesting because they are military personnel, but they are also religious professionals and well-educated about most major religious practices. Many people visit chaplains when they have personal problems. Chaplains are excellent counselors. Recruits should not worry about speaking with chaplains because they have confidentiality privileges. In other words, most of the discussions you have with a chaplain are kept highly private.

All of the chaplains I have met are wise people who have been in the military for many years. Many of the older chaplains are well aware of the problems recruits and active duty personnel face. If you have a personal difficulty that you feel needs professional advice, ask to speak to a chaplain—they usually can help. However, I caution recruits from thinking chaplains can release a recruit from duty or basic training. It is not common that a chaplain can help recruits quit the service.

You will also find military counselors or advisors for nearly every aspect of a recruit's life. There are financial counselors, career counselors and mental health counselors. Relevant to this heading are the mental health counselors. Mental health counselors help recruits understand basic training practices and emotional stress related issues. Usually, counselors have a master's degree in social work (MSW), or a doctorate degree in psychology. Like chaplains, counselors are experts who offer outstanding advice. If you have difficulties beyond what you feel everyone else is having, then make an appointment with these professionals. Sometimes when the mental health specialist cannot help a recruit adjust to military life, they can recommend separation from service.

Military Law—Uniform Code of Military Justice (UCMJ)

When people enlist in the military, they give up many of their civilian rights. The number of legal rights recruits surrender was a big surprise to me. To be fair, the military could not maintain strict order and discipline if soldiers had as much freedom as civilians. During basic training, all trainees will be given a class on military law called UCMJ.

Military lawyers from the Judge Advocate General (JAG) will speak to a large group of recruits about what they can and cannot do while in the military. These lawyers, which are military officers, will compare the differences between civilian law and military law. Needless to say, civilian law affords people (civilians) more legal coverage and protection.

The UCMJ governs all branches of the military. People know when they commit serious crimes (assault, robbery), so I will not address this category of crimes. What recruits need to be aware of during basic is Article 15 of the UCMJ. An Article 15 is given to a recruit for a minor offense called an infraction. Typical Article 15 infractions include not following orders, being absent without leave (AWOL), lying, or showing disrespect to higher-ranking personnel. Therefore, if your platoon sergeant is a grade "A" jackass, do not think you can curse him/her out and quit (of course, people in their right mind would not imagine telling off a sergeant). If you are disrespectful, or do not do what you are told, you will be brought up on Article 15 charges.

Article 15 punishments are administered without a trial or judge. Instead of a judge, a company commander (Captain) hears the case and allows the recruit to seek legal advice. The commander usually explains some of your military rights, and decides if there is enough evidence to dismiss the charges or to find you guilty. If you are found guilty of an Article 15 charge, and you do not want to have a court-martial, your commander will give you a punishment. Punishments can take many forms but are usually a temporary reduction in pay, restriction to barracks, or working extra duties. If you have the choice between an Article 15 and a court-martial, think carefully and listen to your legal counsel. A court-martial is like a jury trial and carries much more severe punishments if you are found guilty.

I remember two incidences in which recruits in my platoon were given, or were at risk for an Article 15. The first incidence happened when a trainee was being "motivated." This particular recruit made the mistake of telling the drilling sergeants that they did not count his sit-ups correctly. To foster an attitude change, the recruit was ordered to do log rolls in the wood chip pit, but he became nauseated after a few minutes. When the recruit stood up, the drill sergeant told him to continue to roll, but the trainee refused. As the defiant recruit stood his ground, three burly sergeants descended upon him. The scene was extremely tense. Then, the recruit was hastily escorted away from the area to have a chat with the first sergeant. In the end, the recruit was threatened with an Article 15, but he did not get one because he requested a formal hearing into the matter—a gutsy move indeed.

The next example involved a much more serious incident. A trainee left a bullet in his weapon. When the recruit returned the weapon to the armory, the round fired and ricocheted off the ceiling. I was very surprised that this recruit was not kicked out of the Army, but instead he received an Article 15. He was ordered to a few days of half-rations (half-food), and a decrease in pay.

Basic Training Records—Do They Last?

The good news about receiving an Article 15 during training is that it will not stay on a permanent record if the recruit shows improvement. In fact, to my understanding, most negative marks during basic training will not be a part of any permanent records. Notice I said most negative marks, not all. For very serious offenses (rape, murder) you will be fully prosecuted and separated from the military.

One of the favorite threats drill sergeants like to employ is making frightened recruits believe they will have negative records if they falter during training. Truth is, your real record starts when you graduate basic. However, since I am not a military lawyer, recruits are advised to speak with a JAG member to understand the latest developments regarding permanent military records.

Quitting the Military

Despite how selective today's military is, once you ship out for basic training the government will try to keep you in its ranks. There are certainly offenses for which recruits are quickly cut from the military, such as drug offenses. However, you will be motivated to stay enlisted because of the thousands of dollars it costs to train each person. In fact, it is unusual to leave the military earlier than your signed contract outlines. Even if you find the military distasteful, your enlistment contract is your required obligation and the government holds you accountable.

On the other hand, I think the military has learned a valuable lesson from the compulsory service days when most people were told they "had" to enlist. Military experts from the 1970's realized that not everyone is military material. This previous statement should not imply that non-military types are weaker, or less capable. After all, some people make excellent doctors, lawyers, or leaders without military service. It takes all types of people to make America work. With this sentiment in mind, the military has made concessions for those individuals who find out the military is not a good fit for them.

There are a few ways to leave the military on positive terms. It is important to note that you do not want a negative separation from the service, such as a dishonorable discharge, because such a discharge will hinder your future employability. One way recruits leave the military is through physical disability. Many times, trainees will have ailments before basic training and these ailments will show themselves during hard exercises. For example, a recruit may have asthma or shortness of breath during a two-mile platoon run. If the problem is discovered to be an existing ailment, the recruit will be discharged with an Existing Prior to Service (EPTS) medical discharge. These types of discharges are not viewed as negative.

Another discharge can happen when the recruit severely injures him/herself during basic training. The most common injuries during training are orthopedic (foot) problems. In fact, one of the reasons the military allows running shoes to be worn is specifically because of the high foot injury rate. Usually, a trainee who suffers from a foot or muscle injury will be allowed time to heal. Once reasonable rest time is given (two to four weeks), the recruit is recycled to start (or finish) training again. In some cases, the injury does not heal and the recruit is medically discharged after a formal hearing. If a permanent or recurrent injury is due *because* of basic training, then the person will receive disability benefits in the form of money and medical care.

If you serve in the military fewer than six months, you can be discharged under an Entry Level Separation (ELS). People discharged under an ELS usually do not incur a negative record, despite what your sergeants might say. Typically, recruits having great difficulty adjusting to the military life will receive an ELS. To illustrate, if a recruit cannot perform duties because he/she cannot legitimately handle stress and is unable to cope with basic, he/she will be discharged with an ELS. If you have drugs in your system, constantly try to sneak away from base, or steal you could receive an ELS depending on the severity of your actions. If a recruit receives an ELS, it will take anywhere from two to seven weeks to be shipped home—which makes for a highly awkward stay at the base.

My perspective on ELS is that the military is not in business to destroy civilian lives and therefore makes concessions for some recruits having difficulty. Some people simply cannot cope well and must be discharged because they will hurt unit morale and efficiency. While a few trainees might be immature, many people just are not the military type. The good news is that recruits who receive an ELS can re-enlist after a two-year waiting period. Note, there are no benefits associated with an ELS.

The next type of discharge is the hardship discharge. Recruits whose family will suffer due to their absence might be eligible for the hardship discharge. For example, if your spouse is extremely ill with cancer, you can be discharged. Another example is if a family member is severely ill and needs care. However, formal evidence of a hardship must be provided and such evidence is scrutinized by commanders. If you feel you are eligible for a hardship discharge, contact the American Red Cross, or the battalion social worker.

The last type of discharge regards homosexuality. In recent years, legislation was passed allowing gays in the military to drop from service with no negative marks on their record. There are probably more homosexuals in the military than some people might like to admit; however if those homosexuals make their preferences known, they can be released from their service. If a trainee approaches a drill sergeant and states he/she cannot perform his/her duties because of homosexual tendencies, then that trainee can be discharged under a Chapter 15. However, I caution recruits who take this road out of the service because the investigation can be very uncomfortable.

In short, trainees are advised to seek an honorable discharge when possible. A dishonorable discharge is not common during basic, but you should do what is necessary to avoid this type of discharge. The reader should study the details of what necessitates a dishonorable discharge because drill sergeants can use it to threaten you, when in fact you might not be in such jeopardy.

Summary

1. If you are legitimately ill or hurt you have a right to have medical attention. Go on sick call if you are hurting, despite being encouraged to "suck it up." Nothing is worth the risk of permanent injury.
2. Trainees are given about an hour per week to practice their religion. This religious time can be a relaxing time to meditate and alleviate the stress of training.
3. Recruits in need of assistance should visit highly trained counselors and psychologists who can help with personal adjustment problems.
4. All military personnel give up some of their civilian rights. The Uniform Code of Military Justice (UCMJ) governs mili-

tary personnel. Military lawyers from JAG can advise you of your legal rights.

5. Many negative marks on a recruit's record are erased after basic training. Sometimes sergeants will use the threat of permanent records against recruits. Although some marks during basic can be enduring, most are not. Understand your rights and records before you enlist.
6. Quitting the military is not easy. Think carefully before you enlist. However, there are legitimate ways to leave the military with no negative marks on your record.

Chapter 9

Miscellaneous Ramblings:
Differences, Mixed Training and Special Notes

Differences in Training—Is Basic Training Easier Today?

If you want to see a heated debate, ask a person who just graduated basic training to speak with a person who graduated 10 years ago. Chances are the 10-year-graduate will argue that his/her basic training was the worst ever. Basic training was much harder 10 years ago than today—right? The irony is that most recruits believe their experiences are harder than the previous group of trainees. This phenomenon is called the fallacy of subjectivity. To add fuel to the fire, the news media sometimes reports how "soft" the military is today. The truth is, basic training today is tough and very demanding. It can be argued that current soldiers are required to think more than previous generations due to the advances in technology. In any case, do <u>not</u> believe sources who might say basic is an easy summer camp. Basic training is still hard, but worthwhile.

In my opinion, some non-combat training sites (for non-combat jobs) might have slightly easier training. Typically, non-combat training bases have mixed (females and males) platoons. Co-ed training entails its own set of problems. Some recruits might see "stress-cards" used in their training cycle. A stress card is a colored card held up to a sergeant when a recruit is too stressed to cope and is at risk of quitting training. I do not believe the stress card practice is widespread, or used for combat MOS trainees. Indeed, if a person enlists to be in direct combat, their training will be more hard-core and stressful. However, if a person enlists to be an office technician (clerk), it is not reasonable to train him/her with combat levels of stress. Despite what many people think, support personnel are crucial to the military and the government does not want to scare potential recruits off with an unduly harsh basic training.

The following Army bases have somewhat tougher basic training cycles: Ft. Benning (Infantry), Ft. Sill (Artillery), Ft. Knox (Armor). In my view, the Marines have the most demanding boot camp regardless of a person's MOS. Navy and Air Force certainly have professional trainings, but are less physically demanding.

Mixing Women and Men

There are a few posts which train men and woman together. Co-ed training is typical for support personnel recruits. Co-ed training posts are currently:

1. Ft. Leonard Woods, MO.
2. Ft. Jackson, SC.,
3. Ft. Sill, OK.,

Some mixed-gender training sites are slowly implementing women into indirect combat jobs such as combat engineers (infantry and armor are still closed to women). Most women enlist for a support MOS. Although considered easier than combat training, co-ed training has problems which might make for a difficult training cycle.

Do not think woman in the military are weak. Biologically speaking, women have certain advantages that men do not, like a higher tolerance to illness and an ability to multitask efficiently. The problems during training arise mainly from distractions between the sexes. It was believed that men would not form unit cohesion, or esprit de corps, due to the presence of women. Men would "show off" in front of women instead of doing what is good for their platoon and team. Additionally, sexual harassment toward females is always an issue. If there were women in my training cycle (which there weren't), the drill instructors would not have been able to speak because every sentence was laced with an incredibly creative curse word. However, do not think that all of the negatives are only against the males. Co-ed trainings have negative consequences for women in the form of more physical injuries when compared to non-gender integrated basic trainings.

On a positive note, various countries who have assimilated females into the military have been highly successful. For example, the Israeli army has an exceedingly tough reputation and it has a large percentage of women in its ranks. From my perspective, women have just as much to offer as men in the military. Even though some of the females' roles might be different, they are no less important to an efficient military.

A Special Note to Women:

Nearly everything in this book is applicable to women as well as to men. Some events such as haircuts and dress will be different, but the core content will be relevant to women. If you are part of a gender-integrated basic training, you will find a female drill instructor as part of your assigned cadre of trainers (as required by training doctrine). You might find that the males you train with actually help you by pushing you to reach your potential. Do not expect special breaks or treatment due to your gender. In short, no matter what type of basic training you attend, you will find it challenging in the beginning and rewarding when you graduate.

Additionally, women in gender-integrated trainings have their own separate rooms and latrines. Men and women are *strictly* separated during non-training times.

Miscellaneous Notes Regarding Basic Training and the Military

The following points are miscellaneous notes which are important, but did not fit neatly into a category. Some points are ideas that are repeated for emphasis. Read each bullet carefully because it might answer a question you have, but forgot to think about.

- You can bring photographs of family or loved ones with you to training. Also, you can receive photos in the mail. However, make sure that any picture you have is not indecent, or does not show nudity.
- As mentioned, mail is important to sound mental health. Positive letters are vital to receive. Tell family and friends not to send any boxes of material such as cookies or presents.
- Men get their heads shaved completely, while women do not.
- You will not be allowed to wear eye contact lenses. People who need corrective vision will be issued heavy-duty training glasses to wear throughout training.
- During reception battalion, you will learn how to make your bed using hospital corners. Making your bunk correctly is like shining your boots—there is an emphasis on it. Always make a tight bed.
- Do not try to pack a camera. Have family or friends take pictures with their cameras during graduation.
- At the end of training, get the addresses and phone numbers of people you want to stay in contact after graduation. I tried to secure my friends' permanent addresses and e-mail addresses .
- Washing clothes is a time-consuming task which must be planned carefully. Do not have friends wash clothes for you because items are usually lost. At some point during training, you will be able to have your clothes cleaned by a service, but it might cost you. The cleaning service is well worth the price; I recommend it.
- You certainly cannot have visitors during basic training.
- You will meet all types of people during training—and I do mean *all* types. Some people are extremely squared away and ready for Delta Force, while others can barely put a sentence

together. However, it has been my observation that meeting different people enriches your training experience and widens your view on humanity. Additionally, different people bring different talents to the task at hand.

- You will have time to contact family members, just not right away. During your initial processing, your family will receive a notice of how to contact you. You will be able to use the phone, but only very briefly—two minutes per use. Usually, recruits can quickly call home within three days of arriving at basic training. Afterwards, phone calls to home are rare.
- Memorize the core values (Army) and what they mean: loyalty, duty, respect, selfless service, honor, integrity, and courage.
- You will gain rank quickly until E-3. Usually, an E-2 promotion comes within six months and the next promotion within a year. If you have some college education, you might receive higher rank immediately. If you have a full college degree, you enter the military as an E-4. Higher rank means more money and more responsibility.
- If your family has an emergency, tell them to contact the Red Cross, or your training base's social worker. These people can help recruits with serious situations.
- Give yourself at least two weeks to adjust to the extreme stress of training. It is normal to feel like quitting. Training does become easier after a few weeks. Do not think Phase 1 will last forever. You will have personal time and space in Phase 3.
- You cannot select your roommate, or battle buddy. Your buddy is assigned to you, so make the best of it.
- Never call a drill sergeant "sir," or just "sergeant."
- Recruits are paid twice a month. The money you earn is directly deposited into your bank account. You will receive a Leave and Earning Statement (LES) which is like a paycheck record. The LES has everything you need to know about how much money you made and how much vacation time you have

acquired. You will be amazed by how much money the government takes out for taxes, but you get this money back at the end of the tax year.

- You cannot switch military jobs (MOS). Make sure you enlist for the job and training you really want.
- Normally, drill sergeants cannot touch you—only for safety purposes. However, I have seen drill sergeants roughly handle recruits and try to intimidate them.
- National Guard and Reserve personnel can sometimes split their training into two summers. One summer reservists complete basic training, while the following summer they complete Advance Individual Training (AIT).
- The youngest a person can enlist is 17 years of age. The oldest is 35.
- There are a variety of food choices, so do not worry if you cannot eat certain foods due to allergies, religion, or beliefs. However, you have to make your preferences/needs known to the drill instructors.
- If you feel your rights are violated during training, contact your chaplain.
- Never vent your complaints openly in public. You never know which statements may get back to the sergeants. Always be careful about what you say.
- Even though you earn 2.5 days of vacation time per month, do not think you can take off anytime you want. You have to secure permission, which is not always granted.
- Basic training usually pauses for a couple of weeks during Christmas time. Recruits can stay on base, or fly home during these holidays.
- You will have no real privacy during training. Showers are open and the bathrooms (latrines) are huge. It will feel like you are always within reach of another recruit.

- When recruits train for basic military communication, they will learn on two types of radios, the VRC-12 and the SINCGARS. These sturdy field radios are about the size of a briefcase. Recruits should at least know what type of radio they use because a drill sergeant might ask for the radio's specifications.
- Recruits usually have the most excitement at the live fire ranges, grenade throwing drills, night-fire exercise and the confidence building obstacle courses. I especially enjoyed training when the drill sergeants allowed us to shoot targets using the M-16's automatic (three round burst) mode. If a task is fun for you, dwell on it!

Chapter 10

Summary
Putting It All Together

New recruits should develop a realistic and balanced perspective regarding the military. There are many positive and negative factors to consider before enlisting. Some benefits include developing personal confidence, deep friendships, job security, money for higher education, leadership and an opportunity to learn meaningful job skills which last a lifetime. Additionally, some military jobs, such as combat occupations, are exciting and unforgettable. If you enjoy travel and adventure, the military offers the avenue to realize these goals.

On the other hand, the drawbacks to the military center on personal issues and sacrifices. Once you sign your contract, it is an enforceable obligation that you cannot easily quit. Strict military standards are upheld and military rules are more numerous than in the civilian world. You are not asked to do things, but are told.

Weighing the benefits and drawbacks of military life is a difficult challenge. Enlisting is a serious matter that demands your time and careful thinking. Commitment and accepting obligations are the hallmarks of mature responsible adults. Some new recruits have never really committed to anything other than completing school. Moreover, committing to the military is a life-altering event.

Once you have read this guide from cover to cover and have decided that you can live with the advantages and disadvantages of the military, make a firm decision to enlist. Do not look back once a decision is made. Simply accept your decision and move forward with your life.

The most remarkable piece of advice I received when I was stuck making my decision came from my brother Sean. Sean had served as a combat medic in Korea and had keen insights into military life. He told me that the military is different for every person. Some people love the military, others hate it, but all learn something from it. He said only I could open the door and see what was waiting for me on the other side. Nobody could open the door for me. Finally he quipped, "The military is what *you* make of it. Once you are in, you have a choice to make the best of your experiences and to learn."

Even if you enlist and dislike it, I am certain you can find something positive in your military experience. Every situation, no matter how difficult, can be framed so that a positive factor emerges. If you feel stuck in a negative situation, remember nothing lasts forever. Try to learn from every experience you have and make yourself stronger from it.

If You Do Not Enlist, or Are Discharged

Many people feel ashamed and inadequate when they either decide not to enlist, or are discharged during basic training. The Marines' have an advertisement which teases people, "Do you have what it takes to be one of us?" "Only a select few can be Marines." True, not everyone can be a Marine, or survive its boot camp. However, it could be argued that plenty of Marines are not right for several civilian jobs. Every person is different with different skills and talents. The world needs *all* types of people. You are not inadequate if you are not the military type—it simply makes you different.

While some of my military friends are rightfully proud of their hard work, others sometime carry their pride too far. People who invent vaccines, or who construct amazing computers are no less deserving of high regard because they do not have military service. In fact, most people can find good-paying jobs, adventure and travel without the military. The goals people seek take planning and motivation. Therefore, it is wise to remember that the military is only one way in which to meet your goals. If you are having trouble meeting your life's goals, a career counselor might be able to set you on the right path.

In short, all people deserve respect if they have personal integrity. If the military is not right for you, do not dwell on it. Realize what your true talents are and move on with your life. Work hard on those things which bring you closer to your goals and be happy.

A Brief Summary of This Book's Most Important Themes

1. People should gather as much information regarding the military as possible before enlisting. Read books; visit Internet sites; and speak with military people, career counselors and family. Your decision must be carefully weighed and thought out. In the end, only you can make the final leap.
2. Recruiters are highly trained professionals who can show you the benefits of the military life. However, be careful because recruiters "sell" the advantages of the military and sometimes do not fully describe all the drawbacks unless specifically asked.
3. Write down your most important life goals and assign a positive emotion to them. Keep your goals with you and refer back to them constantly for motivation and to remind you of your purpose in basic training. Having goals and making progress toward them is the secret to life.
4. Know exactly what you want from the military before you enlist. Make absolutely sure that promises made to you by a recruiter are written down in clear language in your enlistment contract. Read your contract carefully before signing it. Do not be rushed when putting your signature on any official document.

5. Know your rights before you start basic training.
6. If you enlist, starting making an honest effort to increase your mental and physical strength. Start a slow, but constant physical training schedule and take it seriously. Do not burn out early, or injure yourself.
7. Well before basic training starts, read and memorize parts of this guide and your IET handbook. Learn simple commands, how to march, how to stand in formation and other common tasks early.
8. If there is one chapter in this book you should read thoroughly, read Chapter 5, "Basic Training Commandments."
9. Give yourself at least two weeks to adjust to the stress of basic training. Being "stressed-out" is normal. The toughest part of training usually ends after a few weeks. Focus on your goals and the fun parts of training.
10. Remember that your worst enemy during basic training is negative thinking. Basic training is a mind game with yourself and with the drill instructors. Let daily frustrations roll off of you and just go with the flow.
11. During Phase 1, do not volunteer, stand out, or show emotions.
12. Remember to concentrate on the positive aspects of basic training. Boot camp is hard, but nothing worthwhile is easy. You can make it through with the proper positive attitude.

A Final Word

Overall, the military suits most people in the short term. The military experience is intense and rewarding for those who have the right attitude. If you are not planning on retiring from the military, then enlist for the shortest time possible. If you are staying in for more than six years, take advantage of all of the benefits such as education, various training schools and travel. Do whatever it takes (within reason) to get promoted and take the initiative when you are assigned to your active duty post. In other words, make the absolute most of your military experience and let it enrich your life. Be proud of your accomplishments, always pursue new personal objectives and be happy. I wish you a sincere GOOD LUCK in your new life!

--Appendix--
Additional Information and What You Should Know From Your Initial Entry Training Book

I cannot take full credit for this section because some of the information comes directly from the Initial Entry Training (IET) book you will be issued. However, for those readers who cannot access an IET book, or who want to get a head start, this section will prove valuable. (Note—the IET book is a free government publication open to the public domain and full credit is given to it as a source for this section).

Common Tasks

Most basic trainings will be slightly different. However, all recruits are required to pass specific tasks in order to graduate. If you fail an event, you will be required to repeat it until you pass. Failure to complete a task is rarely an option. The following task/activities are core components of most training cycles:

1. Pass a physical fitness test with a minimum score in each category—running, push-ups, sit-ups.
2. Finish an 8-12 mile road march.
3. Pass Chemical/Biological/Nuclear training—Gas chamber.
4. Navigate two obstacle courses.
5. Qualify with the M-16 (28 out of 40 target hits).
6. Complete live fire exercises.
7. Throw grenades.
8. *Complete performance tests after each phase (Phase Testing).
9. Complete hand-to-hand combat.
10. Execute drill and ceremony commands.
11. Finish the assault course.

12. Fully understand the core values.

*Phase testing includes the performance of several other items such as understanding the military's rank structure, protocols and simple common tasks.

Items You Need to Memorize

The following items are some things you will need to commit to memory. You should be able to repeat these items without hesitation *before* basic training begins. Do not just be familiar with this section, memorize it!

Core Values

Loyalty, duty, respect, selfless service, honor, integrity and personal courage. To help you remember these values, the first letter of each value put together spells ("LDRSHIP"--- Leadership)

Officers

Rank

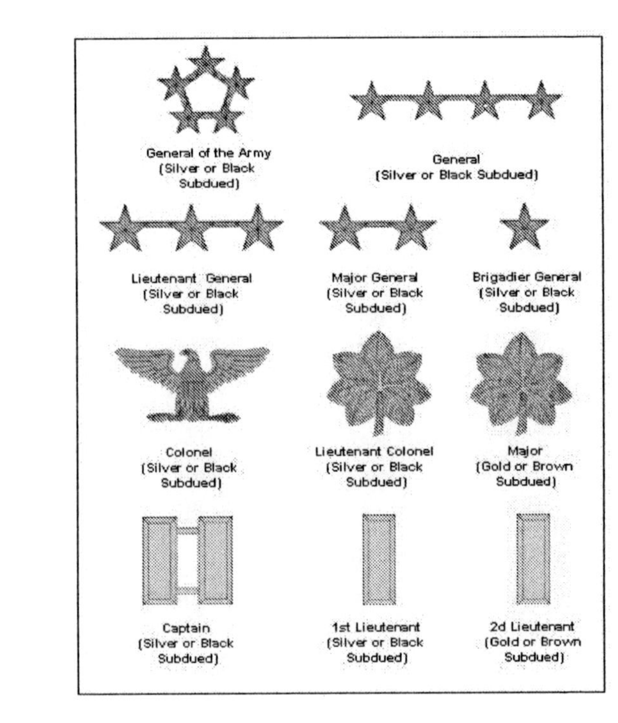

Rank Continued

Enlisted Ranks

Tip: You should know the ranks in order—A First Sergeant outranks a Staff Sergeant. Also, you may be asked to describe the rank insignia. For example, describe the Sergeant Major insignia.

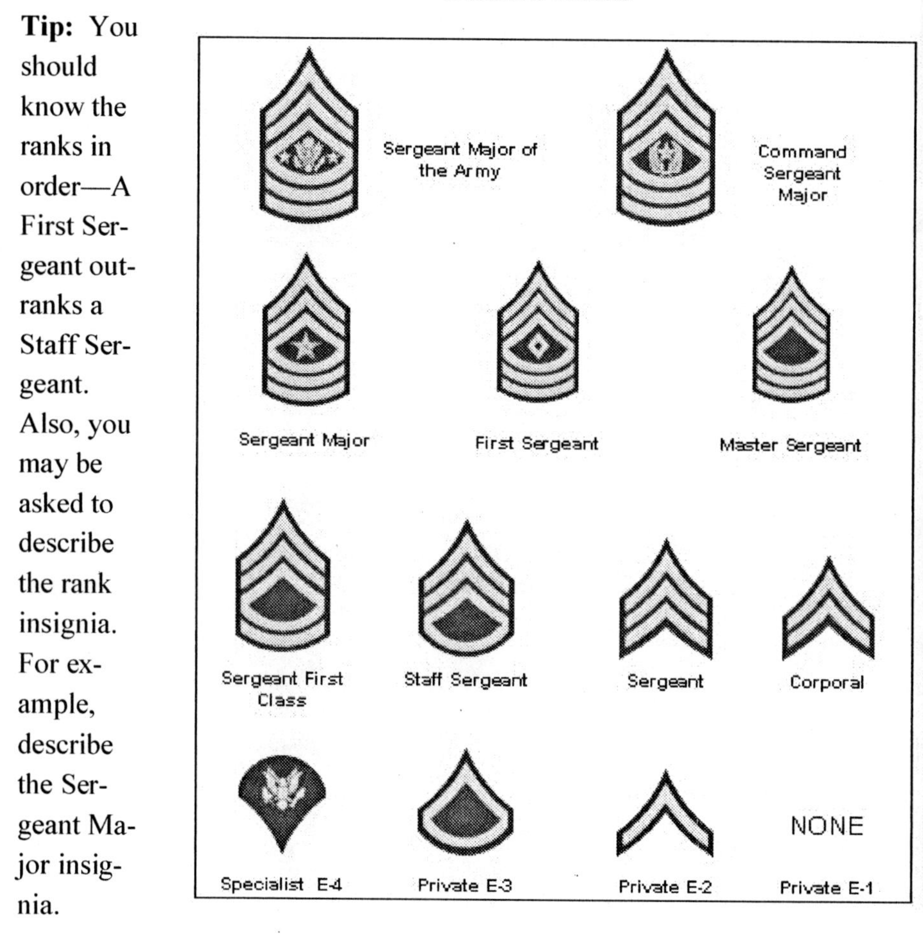

Inspection Arms

Recruits during Phase 2 will need to learn how to perform the command, "Inspection Arms." This task cannot be demonstrated well with pictures, so learn it early when issued your weapon. Most soldiers have difficulty with this task because it involves not looking at your hand movements.

Military Alphabet and Numbers

Phonetic alphabet

A Alpha (al fah)	**B** Bravo (brah voh)	**C** Charlie (char lee)	**D** Delta (dell tah)
E Echo (eck oh)	**F** Foxtrot (foks trot)	**G** Golf (golf)	**H** Hotel (hoh tell)
I India (indee ah)	**J** Juliet (jewlee ett)	**K** Kilo (key loh)	**L** Lima (lee mah)
M Mike (mike)	**N** November (no vem ber)	**O** Oscar (oss cah)	**P** Papa (pah pah)
Q Quebec (keh beck)	**R** Romeo (row me oh)	**S** Sierra (see air rah)	**T** Tango (tang go)
U Uniform (you nee form)	**V** Victor (vik tah)	**W** Whiskey (wiss key)	**X** Xray (ecks ray)
Y Yankee (yang key)	**Z** Zulu (zoo loo)	**1** One (wun)	**2** Two (too)
3 Three (tree)	**4** Four (fow er)	**5** Five (fife)	**6** Six (six)
7 Seven (seven)	**8** Eight (ait)	**9** Nine (niner)	**0** Zero (ze ro)

***Tip: Note how 3 is pronounced "Tree," 5 is "Fife," and 9 is "Niner."**

Military Time

CIVILIAN TIME	MILITARY TIME	CIVILIAN TIME	MILITARY TIME
12:01 am	0001		
1:00 am	0100	1:00 pm	1300
2:00 am	0200	2:00 pm	1400
1:00 am	0300	3:00 pm	1500
4:00 am	0400	4:00 pm	1600
5:00 am	0500	5:00 pm	1700
6:00 am	0600	6:00 pm	1800
7:00 am	0700	7:00 pm	1900
8:00 am	0800	8:00 pm	2000
9:00 am	0900	9:00 pm	2100
10:00 am	1000	10:00 pm	2200
11:00 am	1100	11:00 pm	2300
12:00 NOON	1200	12:00 MIDNIGHT	2400

Military services tell time by using the numbers "1" to "24" for the 24 hours in a day. A day begins at one minute after midnight and ends at midnight the same day. For example, nine minutes after midnight (12:09 am) is written in military time as "0009."

Tip: To convert civilian time to military time understand this trick: After 12:00 noon, add one for each additional hour. For example, 1:00 in the afternoon is 12 :00+1=13:00 (pronounced thirteen hundred hours in military time.) For 2:00 p.m., 12:00 + 2= 14:00 (fourteen hundred hours).

If the time is before noon, it is much like regular time but with a zero in front of the time. For example, 3:00 in the morning is 0300 (pronounced zero-three-hundred hours).

General Orders

Different branches may have different general orders. Below are the Army's general orders. Regardless of your branch of service, memorize your general orders.

General Order Number 1—	*"I will guard everything within the limits of my post and quit my post only when properly relieved."*
General Order Number 2—	*"I will obey my special orders and perform all my duties in a military manner."*
General Order Number 3—	*"I will report violations of my special orders, emergencies, and anything not covered in my instructions to the commander of the relief."*

Saluting

Saluting is a way to show respect to higher ranking people, or to the flag. Only salute officers, not sergeants. Also, only salute when you are outside of a building, not inside. Your salutes should always be sharp and quick.

Standing in Formation

When standing at the position of attention, your hands should be close to your sides with fingers curled up into a loose fist. Head and eyes should be firmly fixed straight and level. Never move in the "attention" position.

"Stand at Ease" is a position used frequently when tasks are being taught to a formation. You will also hear the order to "at ease" when a sergeant enters the room. Stand at ease is also similar to the position of "Parade Rest."

Standing Positions

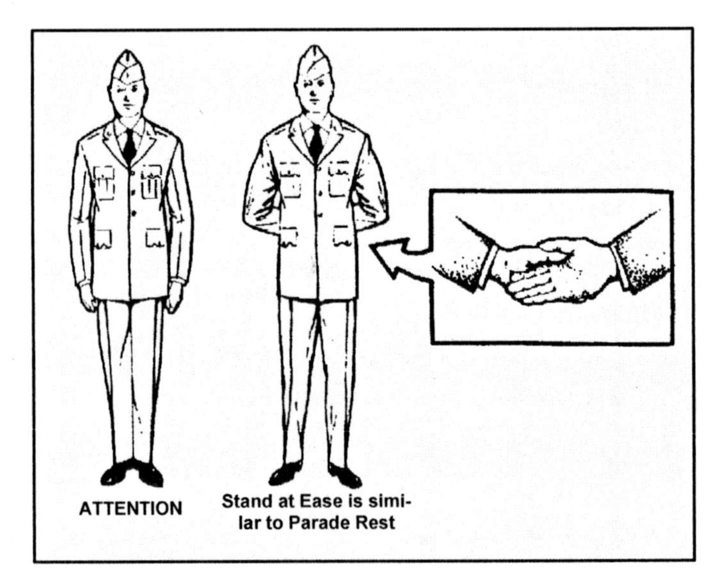

Basic Marching Commands

You must learn how to stand properly in formation and how to march. A large part of "drilling" is to understand all of the commands. Only the basic commands are given here; you will learn more at reception battalion. Tip: Always start marching on your left foot.

Forward March. The command to march with a 30-inch step from the halt is *"Forward, march."* When you hear the *"Forward,"* shift your weight to the right leg without noticeable movement. On the command of execution, *"March,"* step forward with your left foot and continue marching with 30-inch steps keeping your head and eyes forward. Your arms should swing in a natural motion, without bending at the elbows, about nine inches straight forward and six inches to the rear of the seams of your trousers or skirt. Keep your fingers curled as in the position of attention.

Change Step. The command *"Change step, march"* is given as the right foot strikes the ground. When you hear *"March,"* take one more step with your left foot. Then, in one count, place your right toe near the heel of your left foot and step off again with your left foot. You should execute this movement without command if you find yourself out of step with other members in your formation.

Half Step. The command for marching with a 15-inch step is *"Half step, march."* This command may be given from the halt, or it may be given to change a 30-inch step to a 15-inch step while marching. If the command is given from the halt, proceed as for marching with the 30-inch step, but take only 15-inch steps. If the command is given while you are marching with 30-inch steps, you will hear the command *"Half Step March."* Take one more 30-inch step and then begin marching with 15-inch steps. Allow your arms to swing naturally.

Halt. To halt while marching, the command *"Squad/ platoon, halt"* is given as either foot strikes the ground. The movement is executed in two counts. On hearing *"Halt,"* take one more step and then bring your trailing foot alongside your leading foot, resuming the position of attention.

Marching in Place. To march in place, the command is *"Mark time, march."* When you hear *"March,"* take one more step, bring the trailing foot alongside your leading foot, and begin marching in place. To do this, raise each foot alternately 2 inches off the ground. Your arms should continue to swing naturally.

Right/Left. To march with a 15-inch step right or left, you must begin from the halt. To march right, the command is *"Right step, march."* On hearing *"March,"* bend your right knee slightly, and raise your right leg only high enough to allow freedom of movement. Place your right foot 15 inches to the right of your left foot, then move the left foot (keeping your left knee straight) alongside the right foot temporarily assuming the position of attention. To march to the left, go through the same movements in response to the command *"Left step, march,"* except reverse the instructions. On hearing *"Halt,"* take one more step with your lead foot, place your trailing foot alongside your lead foot, and resume the position of attention.

Map Features

Recruits do not have to memorize map features until the later phases of training. However, most people forget these features if they study them too early. The following example uses a fist to demonstrate map features. For your test, you will be given a map and told to point out features.

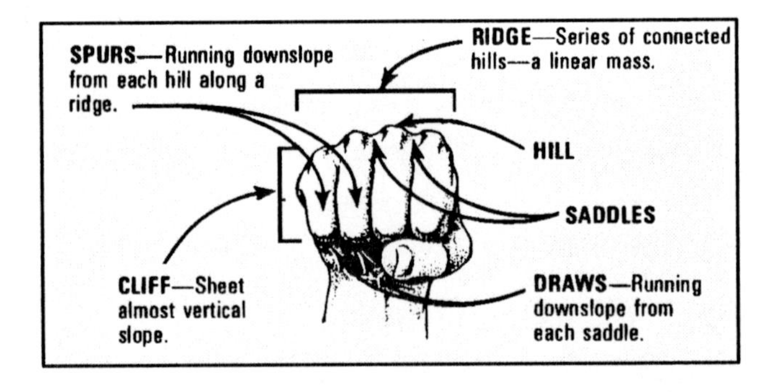

Example Map Feature

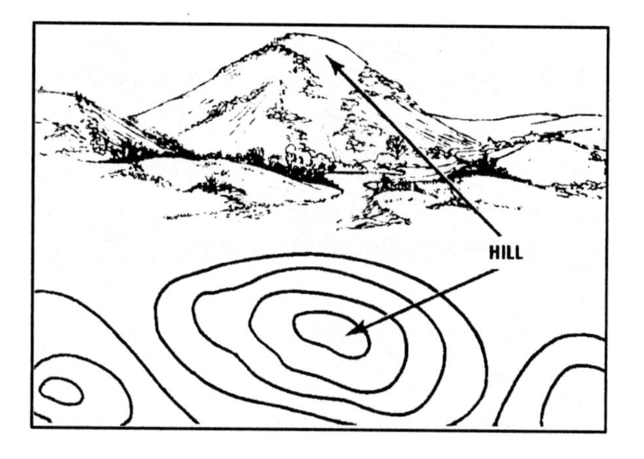

Memorize these map terms: Hill, saddle, spur, ridge, cliff, valley and draw.

Familiarize

The following items should be very familiar to you. Although it is beneficial to memorize as much as possible, usually time does not permit it. Therefore you should find out early what items your specific training site emphasizes and focus accordingly. Some of the following points might need to be memorized, but usually you just have to understand them.

SPORTS: If your rifle malfunctions, remember S-P-O-R-T-S. This key word will help you remember these actions in sequence: Slap, Pull, Observe, Release, Tap, Shoot.

a. **S**lap upward on the magazine to make sure it is properly seated.

b. **P**ull the charging handle all the way back.

c. **O**bserve the ejection of the case or cartridge. Look into the chamber and check for obstructions.
d. **R**elease the charging handle to feed a new round in the chamber. Do not ride the charging handle.
e. **T**ap the forward assist.
f. **S**hoot. If the rifle still does not fire, inspect it to determine the cause of the stoppage or malfunction and take appropriate remedial action.

Parts of Your Weapon (M-16a2)

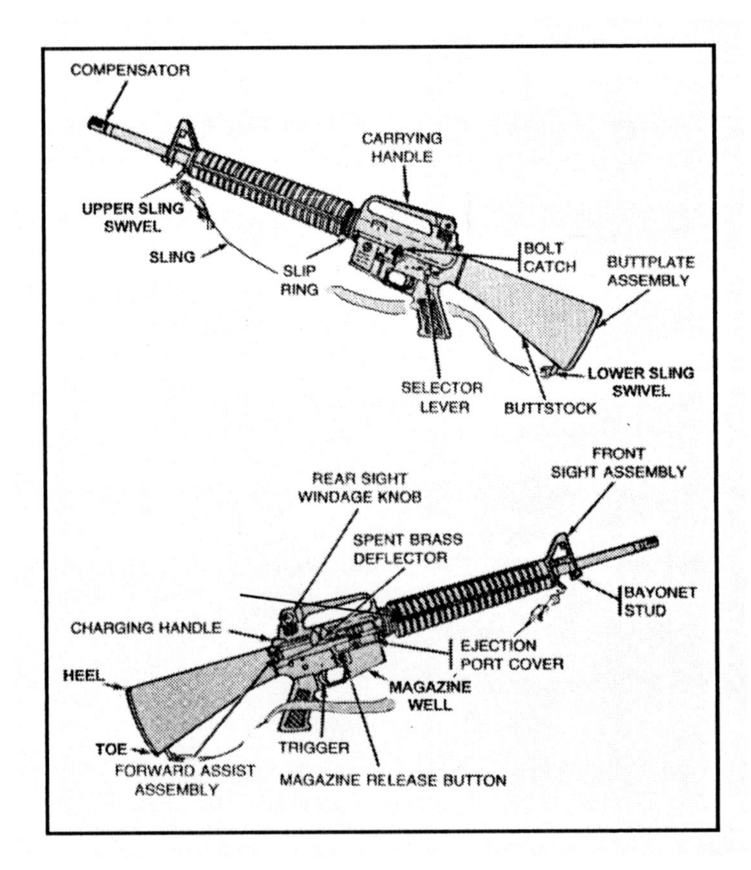

Inspections.

During training you will typically have four inspections. Most recruits do not do well on the first inspection, because the sergeants are picky.

1. The first inspection is a working inspection covering inspection of bunks, wall lockers, individual clothing and equipment. It will take place at the beginning of the cycle and give soldiers an introduction to Army inspection procedures.
2. The second inspection is a stand-by inspection. Items to be inspected include: bunks, wall lockers, individual clothing and equipment.
3. The third inspection period is an in-ranks inspection, and covers open and close ranks, inspection arms, port and order arms, and proper courtesy in an in-ranks inspection.
4. The fourth and final inspection period in basic training is a stand-by inspection. Items to be inspected include: Class A uniforms, weapons, protective masks, bunks, wall lockers, and common areas.

Structure of Groups

During basic training, your primary assigned group is your *platoon* of about 40 people. Your platoon is part of a training *company* which has 4 platoons. You will be most familiar with the platoon and company structure during this part of your military career. However, you should be familiar with how other groups are fitted within the entire military organization. The following units are shown in ascending order of size.

- The **squad** is the smallest unit, consisting of eight to ten soldiers. The squad leader is a Noncommissioned Officer (NCO).

- The **platoon** includes the platoon leader (2LT/1LT), platoon sergeant (SFC) and two or more squads (usually 40-50 people).
- The **company** includes the company commander (CPT), first sergeant (1SG) a headquarters, and two or more platoons (about 150-200 people).
- The **battalion** includes the battalion commander (LTC), his staff and headquarters, the command sergeant major (CSM) and approximately 3-5 companies (600 people +).
- The **brigade** includes the brigade commander (COL), command sergeant major, a headquarters, and approximately 3-6 battalions.
- The **division** structure is the capstone element of our Army. It includes three maneuver (armor or infantry) brigades as well as several combat support and service support brigades or battalions. There are currently 10 active divisions, each commanded by a major general (two-stars):

***Your* Chain of Command:**

Note: When you arrive at reception battalion, quickly learn who are the leaders at your post. Write the names here: *CSM stands for Command Sergeant Major.

Platoon Sergeant________________________________

First Sergeant __________________________________

Company Commander____________________________

*Battalion CSM __________________________________

Brigade CSM ____________________________________

Division/Post CSM ______________________________

TRADOC CSM ___________________________________

Understand the Different Types of Grenades

TYPE	COLOR/MARKINGS	USAGE
M67	OD with yellow marking.	Disables or kills enemy. Explodes 4 to 5 seconds after the safety lever is released.
M18 Colored Smoke	OD with color of smoke on top.	Signals personnel.
M34 WP Smoke	Light green, yellow band, red marking "OLD MARKING" light gray, yellow band, yellow printing.	Signals personnel. Produces casualties up to 35 meters away.
AN-M8 HC Smoke	Light green, black marking, white top.	Screens, "provides concealment."
AN-M14 TH3 Incendiary	Light red with black lettering.	Destroys equipment and starts fires.
ABC-M25A2 CS Riot Control	Gray, red bands and markings.	Controls riots or disables enemy without serious injury.

Basic First Aid

Basic medical assistance techniques are taught early during training and it will be one of your first series of tests. After long lessons on how to treat injuries in the field, you will be tested in small squads. Your IET book has exactly what you need to know, but the following steps will provide you a general idea of what to do. In order, when you approach a victim:

1. Check for responsiveness and send for help; 2. check for breathing: 3. check for bleeding; 4.check for shock; and 5. check for bone fractures.

Military Terms and Acronyms

1. **AIT**—Advance Individual Training. This specialized training comes after a recruit graduates basic training and prepares the individual for his/her military job (MOS).
2. **ALICE Belt**---All-purpose Lightweight Carrying Equipment belt. This combat belt is designed to hold tactical items such as ammunition, canteen, maps, and flashlights.
3. **AO**—Area of Operation. The AO is were a current activity, or engagement is being held. This term is similar to "AA", or assembly area.
4. **AWOL**—Absent Without Leave. A serious offense when a member of the military does not report to his/her duty station in a timely manner. People who try to quit the military without permission are considered AWOL.
5. **BRM**—Basic Rifle Marksmanship. How to maintain and fire your weapon. Marksmanship is actually shooting at targets. All recruits must "qualify" by hitting a specific number of targets with a limited amount of ammunition.
6. **Bivouac**—This is a field activity. The bivouac is designed to introduce recruits to setting up their tents and how to live in the field. Recruits will spend 1, or 2 nights in the field.

7. **Cadence**—This is a marching song. These songs help keep personnel in synchronization (in step) when marching, or running.
8. **CLP**—Cleaner Lubricant Preservative. CLP is a special type of oil used to clean and lubricate weapons.
9. **CO**—Commander Officer. Usually, your immediate commanding officer during training is the company captain. Learn your captain's name.
10. **Conditioning Obstacle Course**—This obstacle course has various obstacles individual soldiers must navigate within a few minutes. Such obstacles might include low wall, ropes, tunnels and monkey bars. The length of the course is roughly the size of a football field.
11. **Confidence Course**—This course is similar to the conditioning course except that obstacles are much larger and requires a team effort to complete.
12. **D and C**—Drill and Ceremony. D and C is the strict military manner of marching, drilling, standing in formation, and other movements your squad or platoon makes.
13. **Fire and Maneuver Course**—This event is sometimes called the live fire exercise. Recruits team up and move through a course of targets in which they actually fire real bullets. The sergeants closely supervise this exercise.
14. **FM**—Field Manual. Field manuals are books and guides published by the military. These manuals are designed to help personnel to understand tasks and machinery used in the military. Some instruction manuals are technical in nature.
15. **FTX**—Field Training Exercise. FTX is like bivouac except it is more militaristic and tactical in nature. Usually soldiers on a FTX will practice shooting and maneuvering through various terrain.

16. **Hand-to-Hand Combat**—Even though the name is obvious, hand-to-hand combat involves specific fighting techniques and how to fall properly.
17. **I.E.T.**—Initial Entry Training. IET is another term for boot camp, or basic training.
18. **Inspections**—Inspections are usually formal observations of a recruit and the platoon. You will probably have three or four inspections-- the first inspection recruits usually fail. Inspections by high-ranking sergeants or officers test the recruit's knowledge about things they learn in basic training. Also, a recruit's personal living area is examined as well as his/her gear.
19. **I.G.**—Inspector General. The I.G.'s office investigates fraud or abuses within the military. Lawyers and investigators typically staff this department. Military personnel can write the IG if they feel their rights have been abused.
20. **JAG**—Judge Advocate General. The JAG is the legal branch of the military. Military personnel are entitled to free legal counsel by the JAG.
21. **K-Pot**—Your bullet resistant Kevlar helmet is called a K-Pot.
22. **LAW**—Light Antitank Weapon—The LAW is a small tube which fires a rocket. Some training sites will expose their recruits to many types of weapons such as the LAW, or AT-4.
23. **MEPS**—Military Processing Entrance Station. MEPS is a government facility where recruits from all branches of the military are processed. Processing entails taking formal tests and medical examinations. MEPS is also the place where you will sign your enlistment contract and leave for basic training. This facility is restricted to government personnel and recruits.
24. **Military Protocols, Customs, and Courtesies**. These terms collectively are used to describe how tasks and procedures are done in the military. Mostly, tasks center on respecting the

rank structure, saluting, reporting for duty, speaking to an officer and how to handle events with the American flag.

25. **MOPP**—Mission Oriented Protective Posture—MOPP is the level of protection necessary to protect personnel from chemical/biological attack. Usually a MOPP suit is a heavy-duty chemical suit personnel wear to protect them from various agents.
26. **MOS**—Mission Occupational Specialty. The MOS is the military job selected by the recruit. Examples are infantryman, tank driver, or medical technician.
27. **NBC**—Nuclear, Biological and Chemical. NBC training teaches recruits how to handle various forms of nuclear, biological and chemical attacks.
28. **NCO**—Non-Commissioned Officer. NCO is a sergeant. There are various ranks of sergeants, but they are all NCO's.
29. **Phase Testing**—At the completion of each phase of basic training (three phases), recruits will be ordered to perform various tasks. Most tasks are simple in nature, but are done under stressful conditions.
30. **PMCS**—Preventative Maintenance Checks and Service. PMCS is any type of work you perform on machinery, or on tools to keep those items working effectively. The military places much emphasis on PMCS. Recruits spend several hours doing PMCS on their weapons and vehicles.
31. **PT**—Physical Training. PT is your daily exercise program. Your level of physical fitness will be measured with a PT test every few months.
32. **Road March**—Typically your platoon, or company will march in a long single file manner with full-combat gear. Marches during basic training range anywhere from one mile to 12 miles.

33. **SALUTE**—Size, Activity, Location, Unit, Time, Equipment. When you radio information regarding an enemy, you will use SALUTE to remember the important details to report.
34. **SOP**—Standard Operating Procedure. SOP is how military tasks are commonly and officially performed.
35. **SPORTS**—Slap, Pull, Observe, Release, Tap, Shoot. SPORTS is an acronym to help recruits remember what to do in case their weapon malfunctions. You will use SPORTS on the firing ranges because every recruit's M-16 will jam at least once.
36. **UCMJ**—Uniform Code of Military Justice. The UCMJ is the legal foundation regulating all of the armed services. Essentially, the UCMJ are military laws and procedures.
37. **UXO**—Unexploded Ordinance. Sometimes bullets, bombs and various munitions fail to detonate, or malfunction. Unexploded munitions should not be handled by recruits and should be reported to the sergeant immediately.
38. **Warrant Officer**—A warrant officer is a special type of officer that commands a rank which is higher than all sergeants (NCOs), but lower than commissioned officers (lieutenants to general). Warrant officers usually have very technical jobs such as helicopter pilots.

Other Information for Personal Use

Internet and Other Resources

As mentioned in the introduction, it is extremely important that the reader utilizes as many sources as possible *before* enlisting in the military. A few good recommended books are: Arco/Baron's ASVAB study guides, *Army Basic Training,* by Raquel Thiebes, and *Boot,* by Daniel DeCruz.

Internet sites are plentiful. I suggest using the search engine Google.com, but there are other excellent engines. About.com and Yahoo search engines have informative discussion groups and chat rooms. Another good idea is to e-mail recruiters with questions. Key terms to perform internet searches are: "military basic training," "basic training," "military instruction," "military boot camp," and "military recruiting." Perform a search using a string of words, or just one word.

Good Army Websites:

- IET Handbook (TRADOC PAMPHLET 600-4)

http://www.adtdl.army.mil/cgi-bin/atdl.dll/pam/600-4/tp600-4.htm

- Military Pay

http://www.dfas.mil/money/milpay/pay/bp-2htm

- U.S. Army Homepage

http://www.army.mil/

- Army Recruiting

http://goarmy.com

- Ft. Benning Homepage (Infantry)

http://www.benning.army.mil/

- Ft. Sill (Artillery)

http://www.sill-www.army.mil

- Ft. Knox (Armor)

http://www.knox-www.army.mil/center/

- Ft. Jackson

http://www.jackson-www.army.mil/

- Ft. Leonard Wood (Engineer, Police)

http://www.wood.army.mil

Marine WebPages:

- www.USMC.mil/
- www.HQMC.USMC.mil/

- http://usmilitary.about.com/cs/marines/

Air Force WebPages:

- www.airforce.com/education/bmt.htm
- www.afreserve.com
- http://usmilitary.about.com/cs/airforce/

Navy WebPages:

- http://usmilitary.about.com/cs/navy/
- www.navy.mil/

Jobs in the military

The military is one of the biggest employers in the world. Most civilian jobs have a counterpart within the military. The Army usually has the widest variety of jobs because it is the largest branch of the service. However, a large number of technical jobs are situated within the Navy and Air Force. Military jobs are organized within 12 large groups. You should take a career test to see which occupational group best addresses your aptitudes and interests. Be absolutely certain which military job (MOS) you want before enlisting. Once you sign your contract, you cannot change your mind.

Below are the major occupational groups. Each group contains several other specific jobs. For example, the Combat group has infantry, armor and artillery nested under it.

1. Combat
2. Health Care
3. Media/Public Relations
4. Engineering/Science
5. Administrative
6. Human Services
7. Service (Law, Supply, Support)
8. Mechanics
9. Electronics
10. Construction
11. Machine Operations
12. Transportation

Example of jobs in the military: X-ray technician, nuclear specialist, firefighter, meteorologist, helicopter pilot, bulldozer operator, bomb specialist, investigator, and journalist.

Pay for Recruits

Every October the military slightly increases the pay schedule for its personnel. The following list illustrates the monthly salary for recruits of typical ranks. Recruits with a college degree enter at the pay rank of E-4. You can also earn rank in various ways before and during training (such as getting a friend to enlist). Ranks above E-4 are competitive and entail graduating from additional training schools.

Below is the approximate monthly salary for a recruit with one year or less of service. You will earn more money if you have more than one year of service, but are at the same pay grade. The pay schedule does not include additional money which might be paid to you for having children, being married or other special factors.

2003 Estimated Monthly Pay:

E-1 (Private)	$1150.83 per month
E-2 (Private)	$1290.11
E-3 (PFC)	$1356.94
E-4 (Specialist)	$1502.79

General Enlistment Information

- Age: Between 17 and 35 years of age.
- Citizenship: Must be a U.S. citizen, or legal registered resident of the U.S.
- Physical: Each service branch has specific minimum and maximum weight and height limits. Weight is typically restricted by height.

- Education: High School degree is favored, but a GED is sometimes accepted. College graduates can enlist with a higher rank (E-4 rank).
- Marital Status: You can be married or single in the military. However, the military may not allow you to enlist if you have outstanding child support obligations.
- Special Considerations: The military can allow certain waivers/grants. Each individual case is examined if you are requesting a waiver of a requirement. For example, if you have a criminal record for a driving offense, you still may be allowed to enlist with special consideration. A first sergeant or captain will interview recruits who request waivers.

About the Author

When I enlisted in the military a few years ago, I was considered older than most recruits. I had a college degree and was working on my graduate degree in psychology from the University of Texas at Austin. Fortunately, my education gave me some insight into the basic training "games." Early in my training, I saw the value of taking careful notes during basic. In fact, one of my goals was to fully document my experiences so I could write this guide. This goal helped me make it through an extraordinarily demanding event in my life.

I enlisted as an E-4 specialist. My MOS was 11 Bravo, or infantry. Due to my MOS, my basic training took place in the hot swampy conditions of Ft. Benning, Georgia. The summer heat in Georgia is oppressive enough without having to participate in an intense activity such as basic training. Although many prior service people say their training was tough, I firmly believe my boot camp was difficult due to my combat MOS. All of my drill sergeants had ranger badges—which meant they were hard-core military types. Moreover, it seemed like our Phase 1 (total control) extended well into Phase 2. Regardless of the crazy times in basic, I look back on my training as an unforgettable experience. I am proud to have served.

After graduating school with my graduate degree, I moved to Colorado. My wife, daughter and dog have been patient while I wrote this guide. Currently, I practice as a psychologist for a large school district. I counsel young adults on a variety of mental health issues and give advice about the military.